Henry Alleyne Nicholson

Synopsis of the Classification of the Animal Kingdom

Henry Alleyne Nicholson

Synopsis of the Classification of the Animal Kingdom

ISBN/EAN: 9783337061777

Printed in Europe, USA, Canada, Australia, Japan

Cover: Foto ©berggeist007 / pixelio.de

More available books at **www.hansebooks.com**

SYNOPSIS

OF THE

CLASSIFICATION OF THE ANIMAL KINGDOM

BY

HENRY ALLEYNE NICHOLSON

MD., D.Sc., Ph.D., F.L.S., F.G.S., Etc.

REGIUS PROFESSOR OF NATURAL HISTORY IN THE
UNIVERSITY OF ABERDEEN

WILLIAM BLACKWOOD AND SONS
EDINBURGH AND LONDON
MDCCCLXXXII

All Rights reserved

PREFACE.

HAVING often been requested by students of Natural History to prepare a synoptical table of the classification of the Animal Kingdom, the present work is an attempt to comply with this requisition, and to supply what the author believes to be a want among zoological workers. Nothing more, of course, could be attempted in the preparation of such a synoptical table than an enumeration of the sub-kingdoms, classes, orders, and sub-orders, with, in general, the *families*, and the principal genera illustrative of these. It has not been possible, however, without unduly extending the limits of the work, to mention in all cases *all* the families, and this deficiency is especially noticeable in the case of such great groups as the Insects, the Fishes, and the Birds. It has also seemed advisable not to give *definitions*, even of the shortest sort, of the subdivisions which are actually enumerated, except in the case of the sub-kingdoms only. It is hardly possible to make such definitions satisfactory within the limits here available; and the introduction of definitions might possibly have led to an abuse of what is really intended to be a mere *guide to a line of study*, and not a thing to be studied in itself.

While definitions have been omitted, a limited number of illustrations have been introduced, as self-explanatory of the text. Occasionally, also, remarks on doubtful points, or divergent views as to classification, are introduced, or alternative arrangements are submitted; and there are added to each group references to some of the sources of special information, which can be studied by advanced students. It seems hardly necessary to add that the purpose of such a classification as is here given, is not that it should be, even in parts, committed to memory, but simply that it may serve as a skeleton, which the student must endow with life by his own work.

MARISCHAL COLLEGE, ABERDEEN,
July 5, 1882.

CONTENTS.

SUB-KINGDOM,—

	PAGE
I. Protozoa,	1
II. Cœlenterata,	15
III. Echinodermata,	27
IV. Annulosa,	36
V. Mollusca,	73
VI. Vertebrata,	86

SYNOPSIS

OF THE

CLASSIFICATION OF THE ANIMAL KINGDOM.

SUB-KINGDOM (TYPE) I.—PROTOZOA.

(Extinct forms are marked with an asterisk.)

ANIMALS composed of undifferentiated protoplasm, or, at most, of protoplasm which is so far differentiated as to have developed a consistent external layer or wall and a central "nucleus" or "endoplast," the organism in the latter case becoming a "cell." In the most typical *Protozoa* the organism remains unicellular, and in no case are definite "tissues" developed by the differentiation of a primitive cellular aggregate. No definite "body-cavity" is in any case developed. There is no recognisable nervous system, and there is either no differentiated alimentary apparatus, or, at most, a rudimentary one.

Most naturalists now divide the animal kingdom into the two primary sections of the *Protozoa* and the *Metazoa;* the former comprising animals which are essentially unicellular, or consist of simple undifferentiated masses of sarcode—while the latter comprises animals which commence their existence as single cells, but which ultimately form cellular aggregates, certain of the cells composing them being differentiated so as to form definite "tissues." The *Protozoa* and *Metazoa* agree, therefore, with one another in the fact that they are, to begin with, simple undivided masses of protoplasm, but they differ in the results produced by the development of this protoplasm. In the case of the *Protozoa*, the original mass of protoplasm may remain undifferentiated, or it may develop a "nucleus," and become thus a "cell;" but it does not become converted into a complex structure composed of metamorphosed cells or "tissues." On the other hand, in the *Metazoa*, the original mass of protoplasm is not only always a true cell, but it becomes converted by a process of regular division into a primitive aggregate of cells, and these secondary cells become finally differentiated into the complex "tissues" of which the body of the adult is composed.

The principal difficulty in the way of accepting this primary, and in the main natural, division of the animal kingdom, is afforded by the Sponges, which are morphologically *Protozoa*, while, according to the views of many naturalists, they are developmentally *Metazoa*. That is to say, they present in their morphological elements so close a resemblance to certain of the *Protozoa* that we can hardly doubt of their close genetic connection with the latter; while, on the other hand, they exhibit in their development (as this has been usually interpreted) the "segmentation" of the primitive ovular cell which is characteristic of the *Metazoa*. Whether or not they possess any definite internal vacuity which can be properly compared with the "body-cavity" of the normal *Metazoa* may still legitimately remain a matter for doubt. It is also still a matter of reasonable doubt whether the development of the Sponges is really properly comparable to that of the *Metazoa*; and, if we accept the views of Mr Saville Kent upon this subject, it certainly is not so. It should also be borne in mind that there are certain of the *Protozoa* (*e.g.*, some of the *Radiolaria*) in which it is not possible to absolutely assert that the adult is unicellular.

As regards the primary divisions of the *Protozoa*, it has not been unusual to accept the presence of a permanent *mouth*, or ingestive aperture, as a good mark of distinction; and, in accordance with this, the *Protozoa* have been divided into the two primary sections of the *Astomata* (comprising the *Gregarinida* and *Rhizopoda*), and the mouth-bearing forms, or *Stomatoda* (comprising the *Infusoria*). Many of the *Infusoria*, however, do not possess a mouth in the proper sense of the term; and Mr Saville Kent has recently ('Manual of Infusoria') proposed the following classification of the *Protozoa*, based upon a more accurate interpretation of the methods in which the ingestion of food is effected by different members of the sub-kingdom:—

SECTION A, PANTÓSTOMATA.—Ingestive area diffuse. This section comprises the *Gregarinida* and the most typical forms of the *Rhizopoda* (viz., the *Amœba*, *Monera*, *Foraminifera*, and *Radiolaria*.)

SECTION B, DISCOSTOMATA.—Ingestive area discoidal, not constituting a distinct mouth. The principal forms included in this section are the Flagellate Infusoria and the Sponges.

SECTION C, EUSTOMATA.—Ingestive area taking the form of a distinct mouth. This section comprises only the Ciliated Infusoria.

SECTION D, POLYSTOMATA.—Ingestive areas distinct and multiple. This section contains only the Suctorial Infusoria.

CLASS I.—GREGARINIDA.

ORDER I.—MONOCYSTIDEA.—*Monocystis* (fig. 1, B).

ORDER II.—EUGREGARINIDA.—*Gregarina* (fig. 1, A and C).

ORDER III.—ACANTHOPHORA.—*Stylorhynchus*.

ORDER IV.—DIDYMOPHYIDA.—*Didymophyes* (perhaps founded upon conjugated forms).

(Kölliker, *Beiträge zur Kenntniss niedere Thiere (die Gattung Gregarina)*, Zeitschr. für Wiss. Zool., 1849; Stein, *Untersuchungen über die Gregarinen*, Archiv für Anat. und Physiol., 1848, and Zeitschr. für Wiss. Zool., 1851; Schneider, *Gregarines des Invertébrés*, Arch. Zool. Exper., 1873 and 1875; Bütschli, *Kleine Beiträge zur Kenntniss der Gregarinen*, Zeitschr. für Wiss. Zool., 1881.)

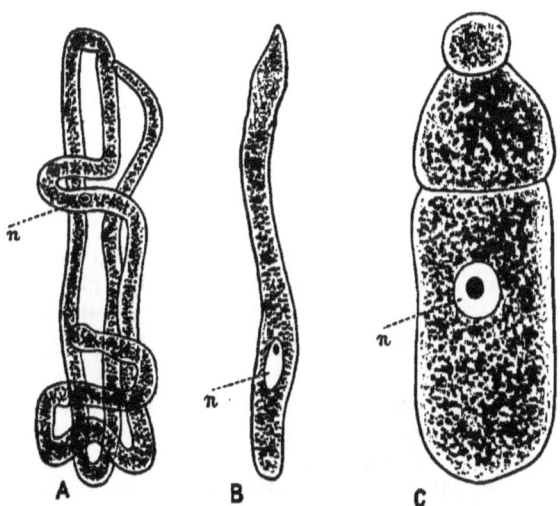

Fig. 1.—Gregarinida. A, *Gregarina gigantea*, parasitic in the Lobster, enlarged, after Van Beneden. B, *Monocystis magna*, parasitic in the Earthworm, enlarged. C, An immature individual of *Gregarina blattarum*, greatly enlarged, after Bütschli, showing the separation of the body into an anterior, middle, and posterior portion: *n*, Nucleus.

CLASS II.—RHIZOPODA.[1]

ORDER I.—MONERA.

Sub-ord. 1. Gymnomonera.—*Protamœba* (fig. 2), *Myxodictyon*.

Sub-ord. 2. Lepomonera.—*Protomyxa, Myxastrum*.

(Hæckel, *Studien über Moneren und andere Protisten*, 1870; Cienkowski, *Beiträge zur Kenntniss der Moneren*, Archiv für Mikros. Anat., 1865.)

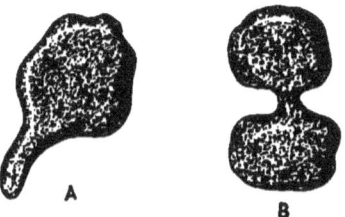

Fig. 2.—Monera. *Protamœba primitiva*, enlarged, after Hæckel. A, An individual with a single pseudopodium protruded; B, Another individual dividing by fission into two portions.

ORDER II.—AMŒBEA (Lobosa).

Sub-ord. 1. Amœbina.—*Amœba* (fig. 3), *Mastigamœba, Pelomyxa*.

Sub-ord. 2. Arcellina.—*Arcella, Difflugia, Hyalosphenia, Quadrula*.

Hertwig and Lesser place the *Arcellina* along with *Gromia* and the *Foraminifera* in a common division, to which they give the name of *Thalamophora*. The *Arcellina*, however, differ from the *Foraminifera*, and agree with the *Amœba* in the blunt lobose character of the pseudopodia.

(Leidy, *Fresh-water Rhizopods of North America*, 1879; Carter, *Fresh-water Rhizopods*, Ann. and Mag. Nat. Hist., 1864; Hertwig and Lesser, *Ueber Rhizopoden und denselben nahe stehende Organismen*, Archiv f. Mikr. Anat., 1874; Frantz Eilhard Schultze, *Rhizopoden-Studien*, Archiv f. Mikr. Anat., 1875.)

---

[1] The Sponges are here removed from the *Rhizopoda*, and are considered as an independent class of the *Protozoa*.

PROTOZOA. 5

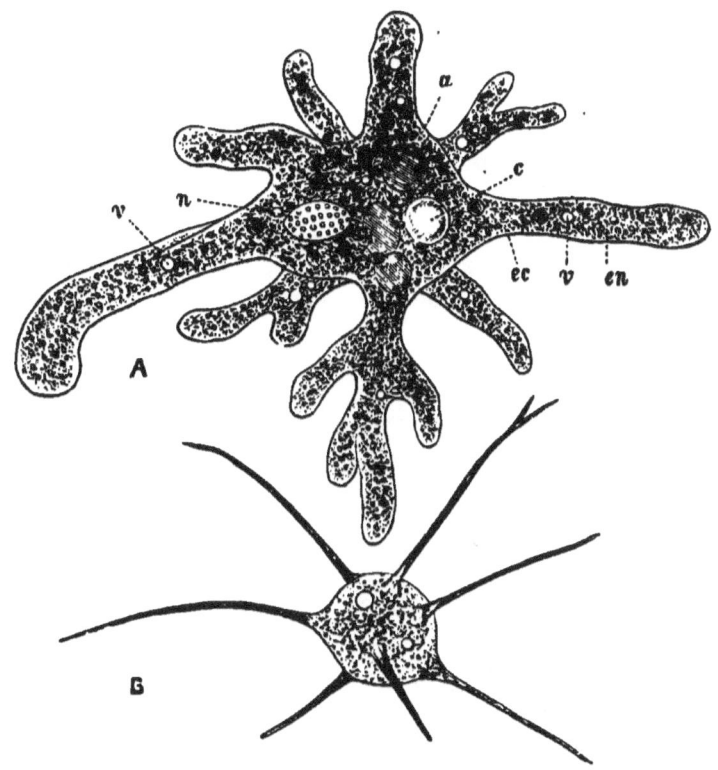

Fig. 3.—A, *Amœba proteus*, with the pseudopodia protruded, enlarged 200 diameters (after Leidy): *n*, Nucleus; *c*, Contractile vesicle; *v*, One of the larger food-vacuoles; *en*, The granular endosarc; *ec*, The transparent ectosarc; *a*, A cell of an Alga taken in as food (other cells of the same Alga are obliquely shaded). B, *Amœba radiosa*, enlarged 500 diameters (after Leidy). The body shows two large vacuoles, but no nucleus or contractile vesicle. The long and delicate pseudopodia are protruded.

ORDER III.—FORAMINIFERA, D'Orb. (RETICULARIA, Carp.)

Sub-ord. 1. Imperforata.

Fam. *a*. Gromidæ.—*Gromia, Microgromia.*
Fam. *b*. Miliolidæ.—*Miliola, Nubecularia, Peneroplis, Orbitolites.*
Fam. *c*. Astrorhizidæ.—*Astrorhiza, Saccammina.*
Fam. *d*. Lituolidæ.—*Lituola, Endothyra, Trochammina.*
Fam. *e*. *Parkeridæ.—*Parkeria, Loftusia.*

Sub-ord. 2. Perforata.

Fam. *a.* Textularidæ.—*Textularia, Bulimina, Cassidulina.*

Fam. *b.* Chilostomellidæ.—*Chilostomella.*

Fam. *c.* Lagenidæ.—*Lagena, Nodosaria, Marginulina, Cristellaria.*

Fam. *d.* Globigerinidæ—*Globigerina.*

Fam. *e.* Rotalidæ.—*Rotalia, Discorbina, Pulvinulina.*

Fam. *f.* Nummulinidæ.—*Nummulites,* **Fusulina, Orbitoides.*

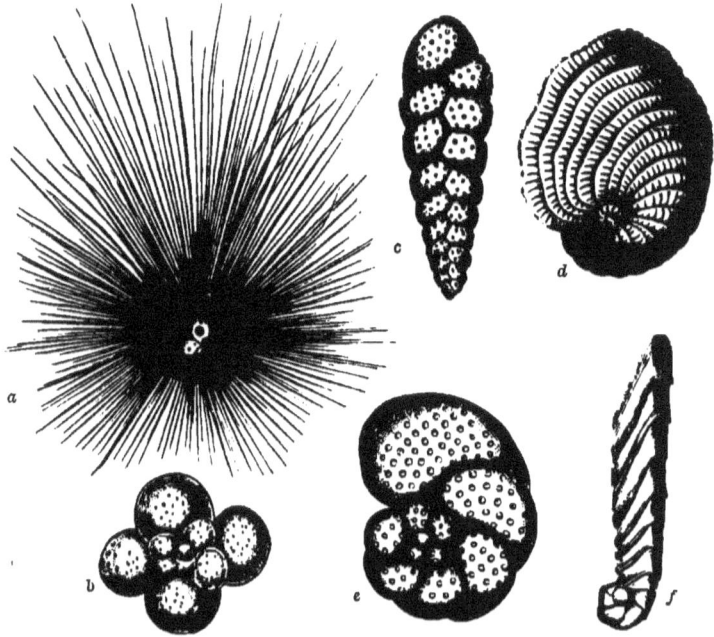

Fig. 4.—Shells of *Foraminifera*. *a, Orbulina universa,* in its perfect condition, showing the tubular spines which radiate from the surface of the shell; *b, Globigerina bulloides,* in its ordinary condition, the thin hollow spines which are attached to the shell when perfect having been broken off; *c, Textularia variabilis; d, Peneroplis planatus; e, Rotalia concamerata; f, Cristellaria subarcuatula.* (Fig. *a* is after Wyville Thomson; the others are after Williamson. All the figures are greatly enlarged.)

The above classification, except in the retention of the sub-orders *Imperforata* and *Perforata,* is that adopted by Mr H. B. Brady. It is true that the section of the *Imperforata* includes various forms in which the shell is known to be pierced to a larger or smaller extent with pseudopodial apertures, so that this division is not a strictly natural one; but the majority of forms included under this title have truly an imperforate shell, and the name is one so long current and so widely used that it seems unadvisable to entirely discard it.

(W. B. Carpenter, Parker, and Rupert Jones, *Introduction to the Study of the Foraminifera*, 1862; Max Schultze, *Ueber den Organismus der Polythalamien*, 1854; H. B. Brady, *Reticularian Rhizopoda of the Challenger Expedition*, Quart. Journ. Micro. Sci., 1879-81; Von Reuss, *Entwurf einer systematischen Zusammenstellung der Foraminiferen*, Sitzungsb. K. Akad. Wiss. Wien, 1861; Hertwig, *Bemerkungen über die Organisation und systematische Stellung der Foraminiferen*, Jenaische Zeitschr. f. Naturwiss., 1876.)

ORDER IV.—RADIOLARIA.

Sub-ord. 1. Cytophora.

Fam. a. Acanthometrina.—*Acanthometra, Xiphacantha* (fig. 5).

Fig. 5.—The skeleton of *Xiphacantha*, one of the *Acanthometrina*, greatly magnified. (After Sir Wyville Thomson.)

8 CLASSIFICATION OF THE ANIMAL KINGDOM.

Fam. b. Polycystina.—*Podocyrtis, Dictyocha, Lychnocanium.*

Fam. c. Collozoa.—*Collozoum, Sphærozoum.*

Fam. d. Thalassicollida.—*Thalassicolla, Thalassolampe.*

Sub-ord. 2. Heliozoa.—*Actinosphærium, Heterophrys.*

It is still a matter of opinion whether the *Heliozoa* should be regarded as a division of the *Radiolaria*, or as a distinct order of Rhizopods. The chief differences which separate the *Heliozoa* from the typical Radiolarians are, that the former possess no central membranous capsule and no gelatinous investment, both these structures being, as a rule, present in the latter; but in other respects there is a close general likeness between the two.

(Hæckel, *Die Radiolarien*, 1862; Schneider, *Zur Kenntniss der Radiolarien*, Zeitschr. f. Wiss. Zool., 1871; Greeff, *Ueber Radiolarien, &c., des süssen Wassers*, Archiv für Mikr. Anat., 1869; Mivart, *Recent Researches on the Radiolaria*, Journ. Linn. Soc., 1878; Archer, *Résumé of Recent Contributions to our Knowledge of Freshwater Rhizopods*, Quart. Journ. Micro. Sci., 1876, 1877; Huxley, *On Thalassicolla*, Ann. Nat. Hist., 1851).

CLASS III.—INFUSORIA.

ORDER I.—FLAGELLATA.—*Monas, Cercomonas, Monosiga* (fig. 6, E), *Codosiga, Euglena, Peridinium, Ceratium* (fig. 6, D).

ORDER II.—SUCTORIA (Tentaculifera, Sav. Kent).—*Podophrya, Acineta.*

ORDER III.—CILIATA.

Sub-ord. 1. Holotricha.—*Paramœcium, Enchelys, Colpoda, Amphileptus* (fig. 6, C).

Sub-ord. 2. Heterotricha.—*Bursaria* (fig. 6, A), *Stentor, Codonella, Nyctotherus* (fig. 6, B).

Sub-ord. 3. Peritricha.—*Trichoderia, Vorticella, Epistylis.*

Sub-ord. 4. Hypotricha.—*Aspidisca, Euplotes, Chilodon.*

There is much ground for separating the so-called Suctorial and Flagellate *Infusoria* as two distinct and independent *classes* of the *Protozoa*. The latter, in particular, have very close affinities with the Sponges, though they are essentially unicellular organisms.

(Saville Kent, *Manual of the Infusoria*, 1880-81; Stein, *Der Organismus der Infusionsthiere*, 1859-67; Claparède et Lachmann, *Études sur les Infusoires et les Rhizopodes*, 1858-61; Allman, *Recent Progress in our Knowledge of the Infusoria*, Journ. Linn. Soc., 1875; Ehrenberg, *Die Infusionsthierchen als volkommene Organismen*, 1838.)

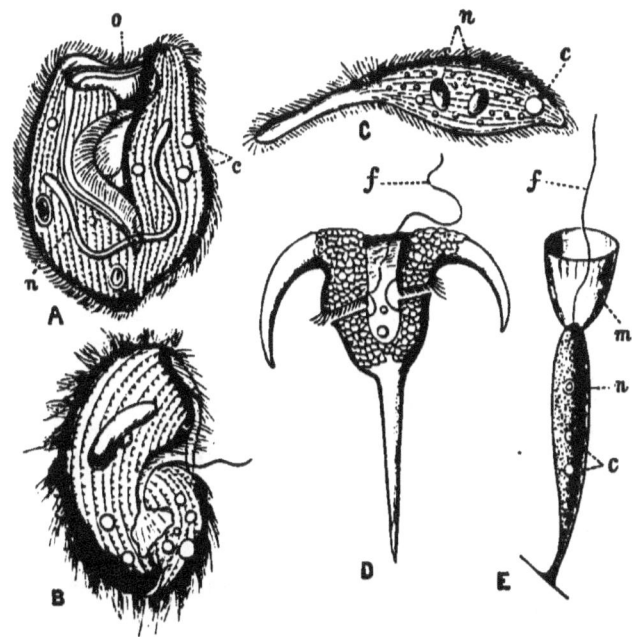

Fig. 6.—Ciliated and Flagellate Infusoria. A, *Bursaria truncatella*, enlarged 50 times. B, *Nyctotherus cordiformis*, enlarged 150 times. C, *Amphileptus anser*, enlarged 120 times. D, *Ceratium tripos*, enlarged 250 times, with its carapace and single flagellum. E, *Monosiga angustata*, enlarged 2500 times: n, Nucleus; c, Contractile vesicle; f, Flagellum; m, Membranous collar surrounding the base of the flagellum. (After, or copied from, Saville Kent.)

CLASS IV.—PORIFERA (SPONGIDA).

ORDER I.—MYXOSPONGIÆ.

Sub-ord. Halisarcidæ.—*Halisarca.*

ORDER II.—CERATOSA (Ceratospongiæ).

Sub-ord. 1. Gumminida.—*Chondrilla, Corticium, Osculina.*
Sub-ord. 2. Ceratina.—*Luffaria, Aplysina.*
Sub-ord. 3. Psammonemata.—*Euspongia, Dysidea.*
Sub-ord. 4. Rhaphidonemata.—*Chalina.*

Sub-ord. 5. Echinonemata.—*Axinella, Acanthella.*

Sub-ord. 6. Holorhaphidota. — *Halichondria, Isodictya, Reniera, Hymeniacidon, Cliona, Geodia, Tethya, Spongilla.*

ORDER III.—SILICEA (Silicispongiæ).

Sub-ord. 1. Lithistidæ.—*Discoderma, MacAndrewia, Corallistes, *Siphonia, *Aulocopium.*

Sub-ord. 2. Hexactinellidæ.—*Euplectella, Holtenia* (fig. 7), *Hyalonema, Dactylocalyx, *Ventriculites.*

ORDER IV.—CALCAREA (Calcispongiæ).—*Grantia, Leucosolenia.*

The most unnatural point in the above classification is the union under the head of "Ceratose Sponges" of types like *Euspongia*, in which spicules are not developed, with other types in which the horny skeleton is accompanied by siliceous spicules, or may be even replaced by the latter. A more natural classification probably is that adopted by Zittel, in accordance with which the Sponges are divided into the following orders :—

ORDER I.—MYXOSPONGIÆ.—*Halisarca.*

ORDER II.—CERATOSPONGIÆ.—*Euspongia* (Sponges of commerce).

ORDER III.—MONACTINELLIDÆ.—*Halichondria.*

ORDER IV.—TETRACTINELLIDÆ.—*Geodia, Tethya.*

ORDER V.—LITHISTIDÆ.—*Discoderma.*

ORDER VI.—HEXACTINELLIDÆ.—*Holtenia.*

ORDER VII.—CALCISPONGIÆ.—*Grantia.*

The systematic position of the Sponges has been a matter of much controversy among naturalists. Their animal nature is now universally admitted, and there is also no substantial difference of opinion as to the broad outlines of their anatomical structure. In all known forms of the Sponges, the organism consists of an aggregate of protoplasmic bodies (the "sponge-particles" or "sarcoids"), which differ in their characters in different parts of the Sponge, or at different periods of its life, but which are probably all fundamentally the same. Some of the sponge-particles precisely and in every respect resemble *Flagellate Infusoria*,

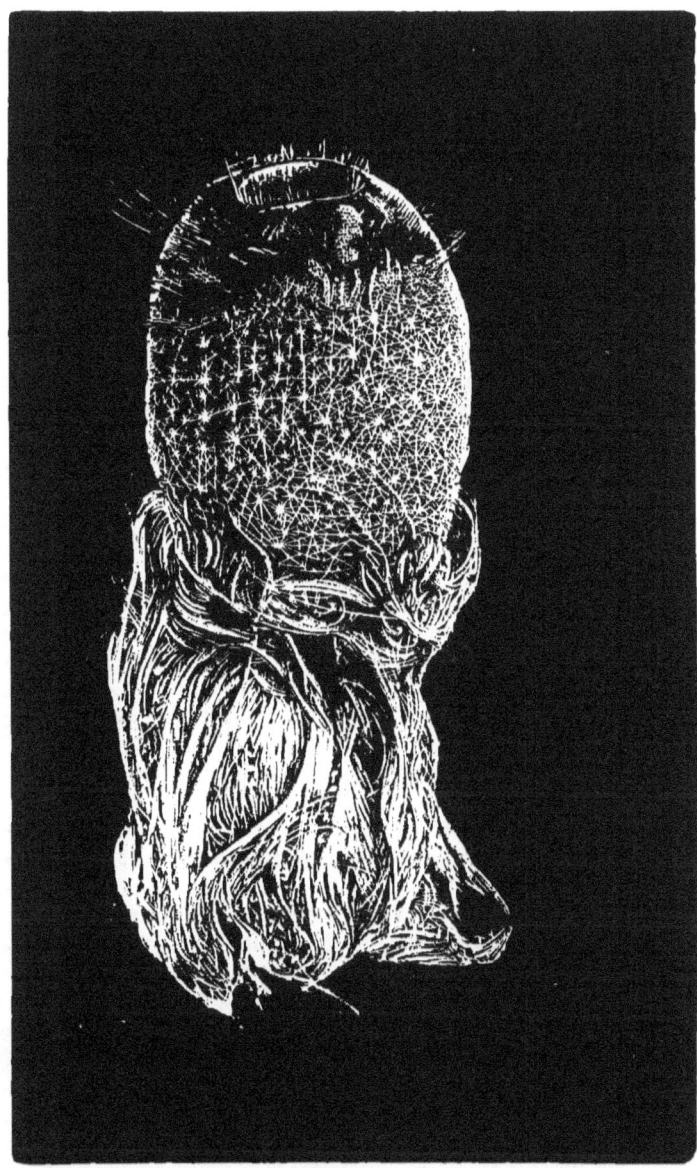

Fig. 7.—*Holtenia Carpenteri*, a siliceous Sponge belonging to the group of the *Hexactinellidæ*. (After Sir Wyville Thomson.)

while others present an equally close resemblance to *Amœbæ;* and others have more or less completely coalesced to form a gelatinous or mucilaginous common matrix or "cytoblastema." This protoplasmic aggregate may or may not be supported by a skeleton of diverse composition and structure; but it is always so disposed as to be traversed by a series of canals, which convey water in and out of the organism, and are connected with respiration and the procuring of food. These canals commence on the surface by numerous small "inhalant" apertures or "pores," which admit the external water, and they ramify through the substance of the Sponge. They ultimately open on the surface by a series of "exhalant" canals, which converge to a common aperture of large size—the so-called "osculum," which serves as an outlet for the water-currents. The entire system of water-canals may be lined with flagellate sponge-particles, similar in structure to Flagellate Infusoria; or they are, more commonly, dilated at intervals into globular chambers, which are lined by these flagellate sarcoids (fig. 8), the vibrations of the flagella of these serving to keep up a circulation of water through the body of the Sponge. Lastly, a Sponge may consist of one excretory opening or "osculum," together with the "pores" belonging to this; or it may consist of a larger or smaller number of such oscula, each with its proper complement of "pores."

Until within the last few years, Sponges have been generally regarded by naturalists as belonging to the *Protozoa,* and as either referable to the *Rhizopoda,* or as constituting a separate division of *Protozoa.* Recently, however, the view has been put forward by Professor Hæckel, and has been largely accepted by zoologists, that the Sponges are properly *Metazoa,* and that they are truly allied to the Corals, and therefore properly referable to the *Cœlenterata.* As a modification of this view, the Sponges are regarded as constituting a group of *Metazoa* intermediate between the Cœlenterates and the *Protozoa.* If Hæckel's view as to the affinities of the Sponges be received, it is necessary to accept the view which this distinguished writer advocates with regard to the *development* of the Sponges, —namely, that the ovum of the *Spongida* undergoes a regular process of "segmentation," consequent on fecundation by a spermatozoön, and that it becomes converted into an embryo ("gastrula") composed of an outer and inner cellular layer, enclosing a central cavity. On the other hand, a large amount of evidence has been brought forward by various observers, and notably by Mr Saville Kent, which would go to show that true sexual reproduction, by means of proper "ova" and "spermatozoa," is of very doubtful occurrence among the Sponges; and that it is very questionable, therefore, if there is truly any such phenomenon in their development as the "segmentation" of an ovular cell. The supposed two-layered "gastrula" of the Sponges would rather appear to be really an asexually produced "swarm-gemmule," composed partly or wholly of flagellate zoöids or monads, entirely similar in their structure to the Flagellate Infusoria, and resulting from the segmentation of a single "sponge-particle" of the adult Sponge, without previous impregnation by a spermatozoid. Upon this view, therefore, the development of the Sponge becomes capable of being strictly paralleled by that of several groups of undoubted *Protozoa*, but acquires a significance entirely different from that which must be ascribed to the development of the *Metazoa*.

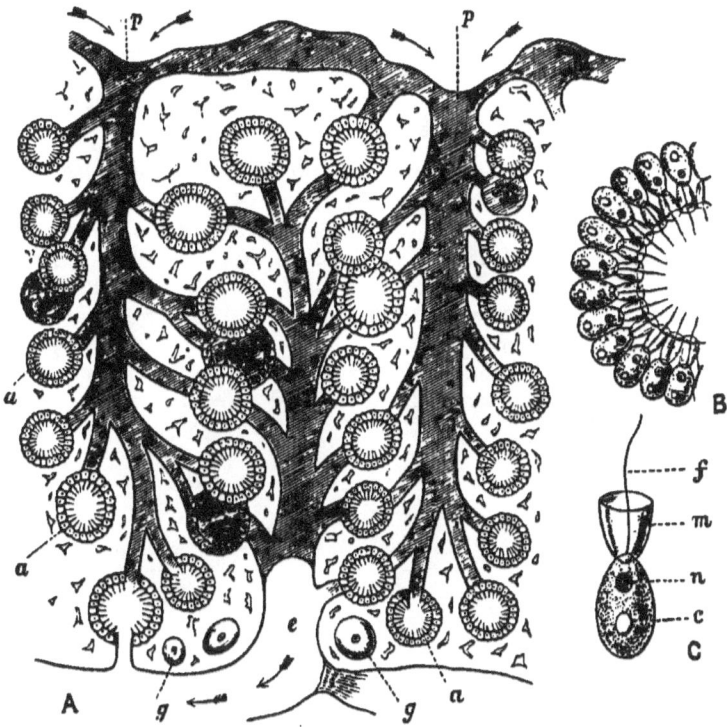

Fig. 8.—Structure of Spongida. A, Vertical section of the outer layer of *Halisarca lobularis*, a Sponge in which the skeleton is wanting, enlarged 75 times (after F. E. Schultze): *p p*, "Pores," or openings of afferent canals by which water is conducted to the ciliated chambers or "ampullaceous sacs" (*a a*); *e*, Commencement of a larger efferent canal, conducting from the ampullaceous sacs to the deeper canals, by which the water is finally carried off to be expelled from the "oscula;" *g g*, Young stages of the reproductive bodies or spores. B, Part of a single ampullaceous sac of the same Sponge, transversely divided, and enlarged 800 diameters (after Saville Kent), showing the flagellate monads or "sponge-particles," with their inwardly directed flagella. c, A single flagellate monad of the same, still further enlarged: *f*, Flagellum; *m*, Collar round the base of the flagellum; *n*, Nucleus; *c*, Contractile vesicle.

Accepting, then, in the meanwhile, the last-mentioned views as to the real character and import of the development of the *Spongida*, it seems inevitable that these organisms must, with our present knowledge, be included among the *Protozoa*. This conclusion, moreover, is the one which is clearly deducible from a study of the structure of the adult organism, since it is certain that there exists an almost absolute identity of structure between the flagellate zoöids of the Sponges and the Flagellate Infusoria, —the reference of the latter to the *Protozoa* having never been called in question. It may be added, that even if it were proved that the Sponges were properly referable to the *Metazoa*, it would still require very much more evidence than has yet been brought forward, before their relationships with the *Cœlenterata* could be reasonably admitted. Upon the whole, therefore, it is probably best—if only as a provisional arrangement—to regard the Sponges as a special division of the *Protozoa*, closely allied to the *Infusoria*, but with sufficient peculiarities of their own to entitle them to a special place and a special name (*Porifera*).

(Oscar Schmidt, *Die Spongien des Adriatischen Meeres*, 1862, 1866; *Grundzüge einer Spongien-fauna des Atlantischen Gebietes*, 1870; and *Spongien des Meerbusen von Mexico*, 1880. Carter, *Notes introductory to the Study and Classification of the Spongida*, Ann. and Mag. Nat. Hist., 1875. Bowerbank, *A Monograph of the British Spongidæ*, 1866, 1874. Hæckel, *Die Kalkschwämme*, 1872. Johnston, *A History of the British Sponges and Lithophytes*, 1842. Saville Kent, *Manual of the Infusoria*, 1880-81. Zittel, *Beiträge zur Systematik fossiler Spongien*, Neues Jahrb. für Min., Geol., und Paleont., 1877-78, and Handbuch der Paleontologie, 1879.)

SUB-KINGDOM (TYPE) II.—CŒLENTERATA.

RADIALLY symmetrical animals, in which the mouth opens into a simple or variously divided space, which represents the alimentary tract, and which may or may not be divided into two portions,—one specially connected with digestion, and the other corresponding with the body-cavity of the higher animals. Body-wall composed of two fundamental layers ("ectoderm" and "endoderm"). Nervous system sometimes specialised, sometimes diffused, but no vascular system developed. Reproductive organs invariably present at some period or another of life, though asexual reproduction is very general.

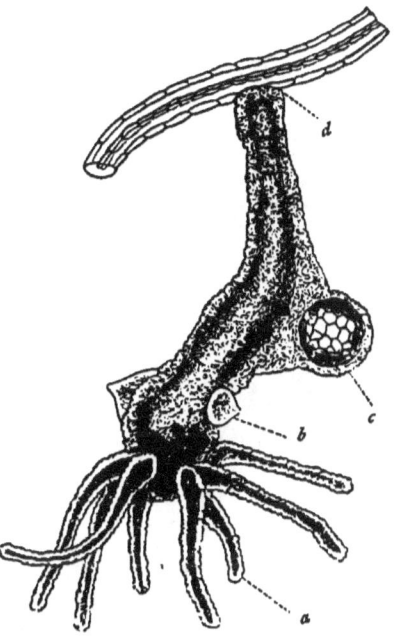

Fig. 9.—The Green Fresh-water Polype (*Hydra viridis*), suspended head-downwards from a piece of a stem of an aquatic plant, enlarged. *a*, One of the tentacles; *b*, Testis or spermarium, with spermatozoa in its interior; *c*, A single large ovum, protruding from the side of the body; *d*, Disc of attachment ("hydrorhiza").

CLASS I.—HYDROZOA.

SUB-CLASS I.—HYDROIDA (Hydroid Zoophytes).

ORDER I.—HYDRIDA.—*Hydra* (Fresh-water Polype, fig. 9).

(Kleinenberg, *Hydra, eine anatomisch-entwickelungsgeschichtliche Untersuchung*, 1872.)

16 CLASSIFICATION OF THE ANIMAL KINGDOM.

ORDER II.—CORYNIDA.—*Coryne, Tubularia, Clava, Bougainvillea* (fig. 10), *Eudendrium, Hydractinia.*

(Allman, *Monograph of the Gymnoblastic or Tubularian Hydroids*, Ray Society, 1871 ; Hincks, *British Hydroid Zoophytes*, 1872 ; Louis Agassiz, *Contributions to the Natural History of the United States*, vols. iii. and iv., 1860-62.)

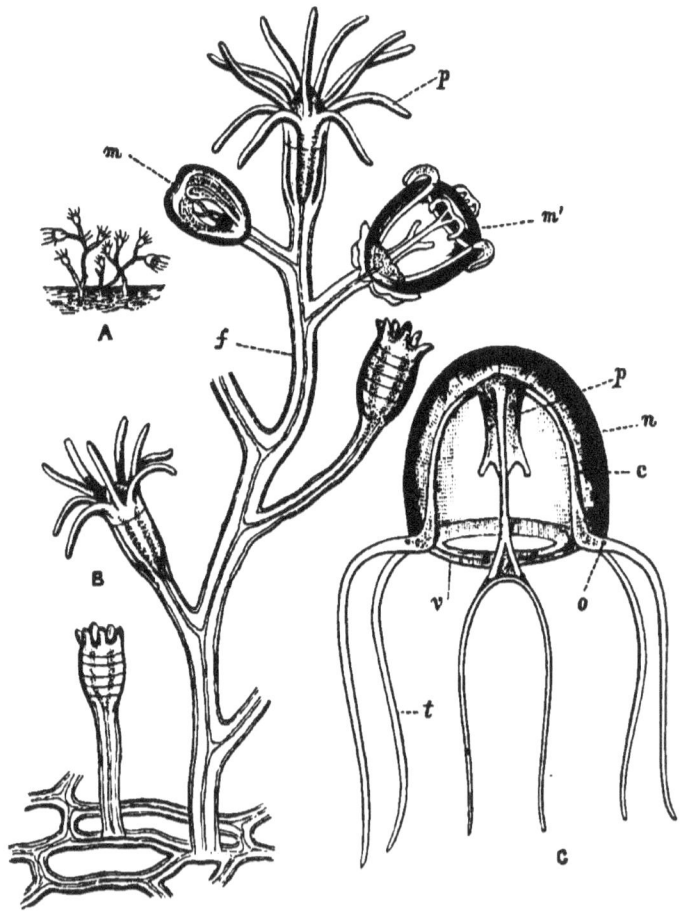

Fig. 10.—Corynida. A, Part of the colony of *Bougainvillea muscus*, of the natural size. B, Part of the same enlarged : *p*, A polypite fully expanded ; *m*, An incompletely developed medusiform bud ; *m'*, A more completely developed medusiform bud ; *f*, Cœnosarc with its investing periderm and central canal. c, A free medusiform gonophore of the same ; *n*, Gonocalyx ; *p*, Manubrium ; *c*, One of the radiating gastro-vascular canals ; *o*, Ocellus ; *v*, Velum ; *t*, Tentacle. (After Allman.)

ORDER III.—SERTULARIDA.—*Sertularia* (Sea - fir), *Diphasia* (fig. 11), *Plumularia, Antennularia.*

ORDER IV.—CAMPANULARIDA.—*Campanularia, Obelia, Clytia, Lafoëa, Thaumantias, Æquorea.*

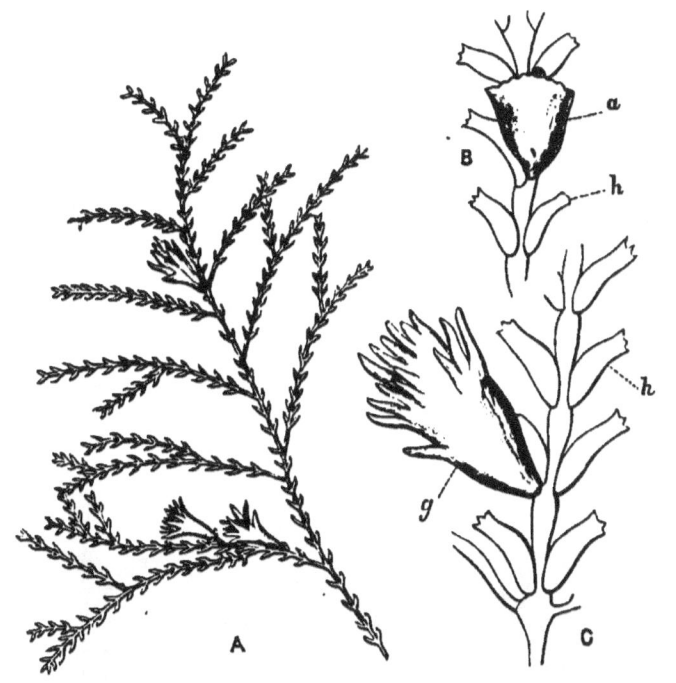

Fig. 11.—Sertularida. A, Portion of the colony of *Diphasia tamarisca*, of the natural size, showing hydrothecæ and female ovarian capsules (gonangia). B and C, Portions of different branches of the same, enlarged : *h*, Hydrothecæ ; *a*, Male gonangium ; *g*, Female gonangium. (After Hincks.)

ORDER V.—THECOMEDUSÆ.—*Stephanoscyphus.*

ORDER VI. — HYDROMEDUSIDÆ (Medusidæ). — *Trachynema* (fig. 12), *Ægina.*

(Hæckel, *Das System der Medusen*, Jena, 1879).
(Huxley, *Monograph of the Oceanic Hydrozoa*, Ray Society, 1859 ; Kölliker, *Die Siphonophoren oder Schwimm-polypen von Messina*, 1853 ; Gegenbaur, *Beobachtungen über Schwimm-polypen*, Zeitschr. für Wiss. Zool., 1854.)

18   CLASSIFICATION OF THE ANIMAL KINGDOM.

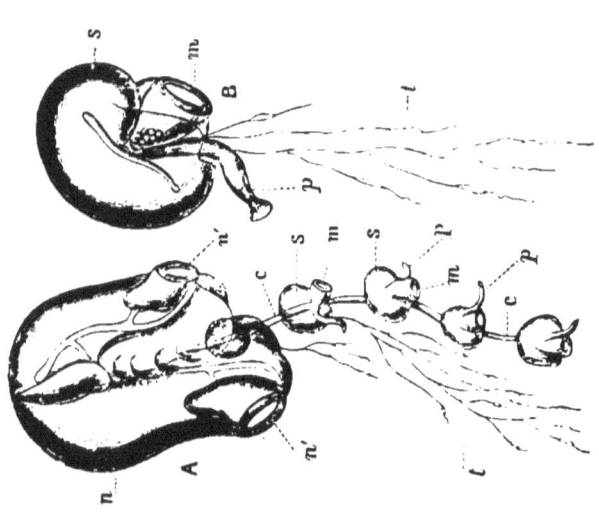

Fig. 13.—Calycophoridæ. A, Upper portion of the colony of *Praya maxima*, of the natural size: *n*, The proximal nectocalyces; *n' n'*, Mouths of the same; *c c*, Cœnosarc, carrying polypites (*p p*) at intervals, along with their swimming-bells (*s s*), the openings of these being indicated by the letters *m m*; *t*, tentacles. B, A single polypite of the same (*p*), separated from the cœnosarc, and enlarged, with its swimming-bell (*s*), the opening of the bell (*m*), and the tentacles (*t*). (After Gegenbaur.)

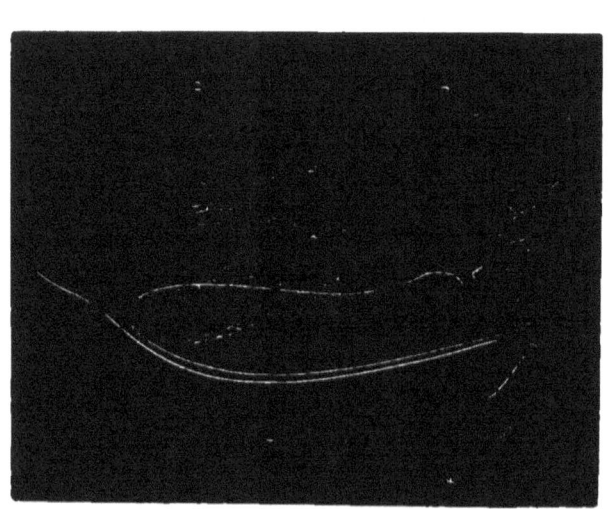

Fig. 12.—*Trachynema digitale*, a naked-eyed Medusa, female, enlarged. (After A. Agassiz.) *p*, Manubrium or central polypite; *t*, One of the tentacles; *c*, One of the gastro-vascular canals; *o*, One of the ovaries.

CŒLENTERATA. 19

SUB-CLASS II.—SIPHONOPHORA.

ORDER I. — CALYCOPHORIDÆ. — *Diphyes, Praya* (fig. 13), *Vogtia.*

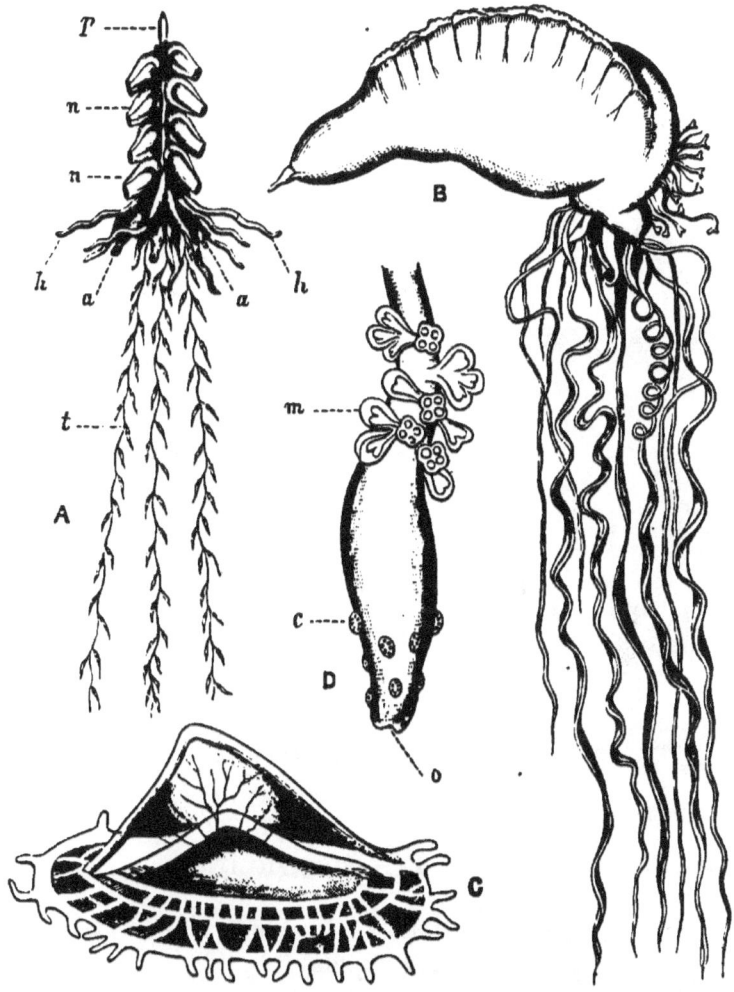

Fig. 14.—Physophoridæ. A, *Physophora Philippi:* p, The pneumatophore; n n, The nectocalyces; h h, Hydrocysts; a a, Polypites; t, Tentacles. B, *Physalia pelagica*. c, *Velella spirans*. D, One of the smaller polypites (phyogemmaria) of the same, showing (o) the mouth, (c) elevations studded with thread-cells, and (m) medusoid buds.

ORDER II.—PHYSOPHORIDÆ.—*Physophora* (fig. 14), *Physalia* (Portuguese Man-of-war, fig. 14), *Velella* (fig. 14 c), *Porpita*, *Agalma*, *Stephanomia*.

SUB-CLASS III.—LUCERNARIDA.

SECTION I.—CALYCOZOA (Podactinaria).
    Order.—LUCERNARIADÆ.—*Lucernaria, Carduella, Depastrum*.

Fig. 15.—Generative zoöid of *Rhizostoma pulmo*, reduced in size. (After Gosse.)

SECTION II.—ACRASPEDA.
    Order I.—MONOSTOMATA.
        Fam. 1. Pelagidæ.—*Pelagia*.
        Fam. 2. Cyaneidæ.—*Cyanea*.
        Fam. 3. Aureliidæ.—*Aurelia*.
    Order II.—RHIZOSTOMATA.
        Fam. 1. Rhizostomidæ.—*Rhizostoma* (fig. 15).
        Fam. 2 Cepheidæ.—*Cephea*.
        Fam. 3. Polyclonidæ.—*Polyclonia*.
        Fam. 4. Cassiopeidæ.—*Cassiopeia*.
        Fam. 5. Crambessidæ.—*Crambessa*.

(Huxley, *On the Anatomy and Affinities of the Family of the Medusidæ*, Phil. Trans., 1849; Brandt, *Ueber Rhizostoma Cuvieri*, Mem. Acad. St Petersbourg, 1870.)

SUB-CLASS IV.—*GRAPTOLITIDÆ.

ORDER I.—MONOPRIONIDÆ. — *Monograptus, Didymograptus, Tetragraptus, Dichograptus*.

ORDER II.—DIPRIONIDÆ.—*Diplograptus, Climacograptus*.

ORDER III.—TETRAPRIONIDÆ.—*Phyllograptus*.

ORDER IV.—RETIOLOIDEA.—*Retiolites*.

CŒLENTERATA. 21

(Hall, *Graptolites of the Quebec Series*, 1865; Nicholson, *Monograph of the British Graptolitidæ*, 1872; Lapworth, *Notes on British Graptolites*, Geol. Mag., 1873.)

SUB-CLASS V.—HYDROCORALLINÆ.

Fam. 1. Milleporidæ.—*Millepora* (fig. 16).
Fam. 2. Stylasteridæ.—*Stylaster, Pliobothrus, Errina, Distichopora.*

(H. N. Moseley, *Report on certain Hydroid, Alcyonarian, and Madreporarian Corals*, Report of the Scientific Results of the Voyage of H.M.S. Challenger, 1881.)

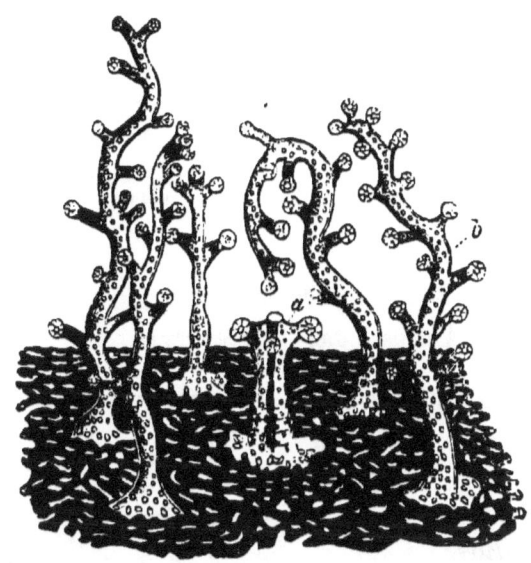

Fig. 16.—Enlarged view of a portion of the surface of a living colony of *Millepora nodosa*, showing the expanded zoöids of a single system. *a*, Central "gastrozoöid;" *b*, One of the mouthless "dactylozoöids." (After Moseley.)

CLASS II.—ACTINOZOA.

ORDER I.—ZOANTHARIA.

Sub-ord. 1. Zoantharia malacodermata.
Fam. *a*. Actinidæ (Sea-anemones).—*Actinia* (fig. 17), *Tealia, Sagartia, Minyas.*

Fam. b. Ilyanthidæ. — *Ilyanthus, Arachnactis, Edwardsia.*

Fam. c. Zoanthidæ.—*Zoanthus, Palythoa.*

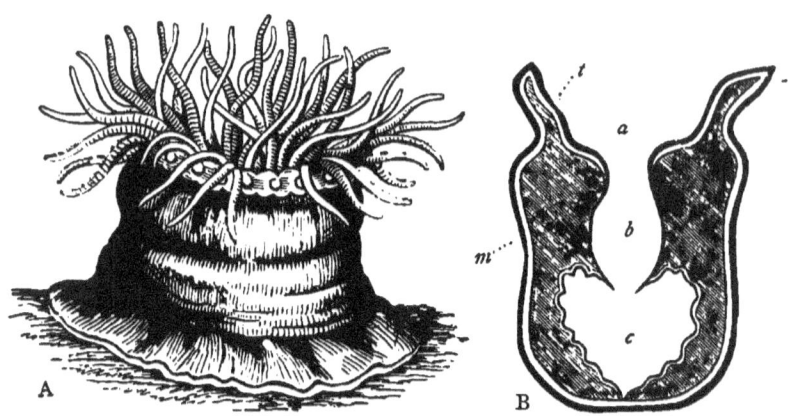

Fig. 17.—A, *Actinia mesembryanthemum*, one of the Sea-anemones (after Johnston). B, Section of the same, showing the mouth (*a*), the stomach (*b*), and the body-cavity (*c*); *t*, Tentacle; *m*, Face of a mesentery.

Sub-ord. 2. Zoantharia sclerobasica.

Fam. a. Antipathidæ.—*Antipathes.*

Sub-ord. 3. Zoantharia sclerodermata (Madreporaria).

Section A. Perforata.

Fam. a. Poritidæ.—*Porites, Alveopora.*
Fam. b. *Favositidæ.—*Favosites, Michelinia, Alveolites.*
Fam. c. *Syringoporidæ.—*Syringopora.*
Fam. d. Eupsammidæ.—*Balanophyllia, Dendrophyllia.*
Fam. e. Madreporidæ.—*Madrepora.*

Section B. Aporosa.

Fam. a. Fungidæ.—*Fungia.*
Fam. b. Pseudofungidæ.—*Merulina.*
Fam. c. Astræidæ.—*Astræa, Meandrina, Diploria.*
Fam. d. *Columnariadæ.—*Columnaria.*

Fam. *c*. Oculinidæ. — *Oculina, Lophohelia, Amphihelia.*
Fam. *f*. Pseudoturbinolidæ.—*Dasmia.*
Fam. *g*. Turbinolidæ.—*Turbinolia, Flabellum.*

(Oscar Hertwig and Richard Hertwig, *Die Actinien*, Jena, 1879; Gosse, *Actinologia Britannica*, 1860; Dana, *Report on Zoophytes*, 1849; Milne-Edwards and Haime, *Histoire Naturelle des Coralliaires*, 1857-60.)

Fig. 18.— *Astræa pallida*, a compound sclerodermic Coral, in a living condition. (After Dana.)

ORDER II.—ALCYONARIA.

Fam. *a*. Alcyonidæ.—*Alcyonium* (Dead-men's Fingers), *Rhizoxenia.*
Fam. *b*. Tubiporidæ.—*Tubipora* (Organ-pipe Coral).
Fam. *c*. Pennatulidæ.— *Pennatula* (Sea-pen, fig. 20), *Virgularia* (Sea-rod), *Veretillum* (fig. 19), *Renilla, Pavonaria.*
Fam. *d*. Gorgonidæ.—*Gorgonia* (Sea-shrub), *Isis, Corallium* (Red Coral), *Rhipidogorgia* (Fan-coral, fig. 21), *Mopsea.*
Fam. *e*. Helioporidæ.—*Heliopora.*
Fam. *f*. *Halysitidæ.—*Halysites.*
Fam. *g*. *Tetradiidæ.—*Tetradium.*
Fam. *h*. *Thecidæ.—*Thecia.*

24 CLASSIFICATION OF THE ANIMAL KINGDOM.

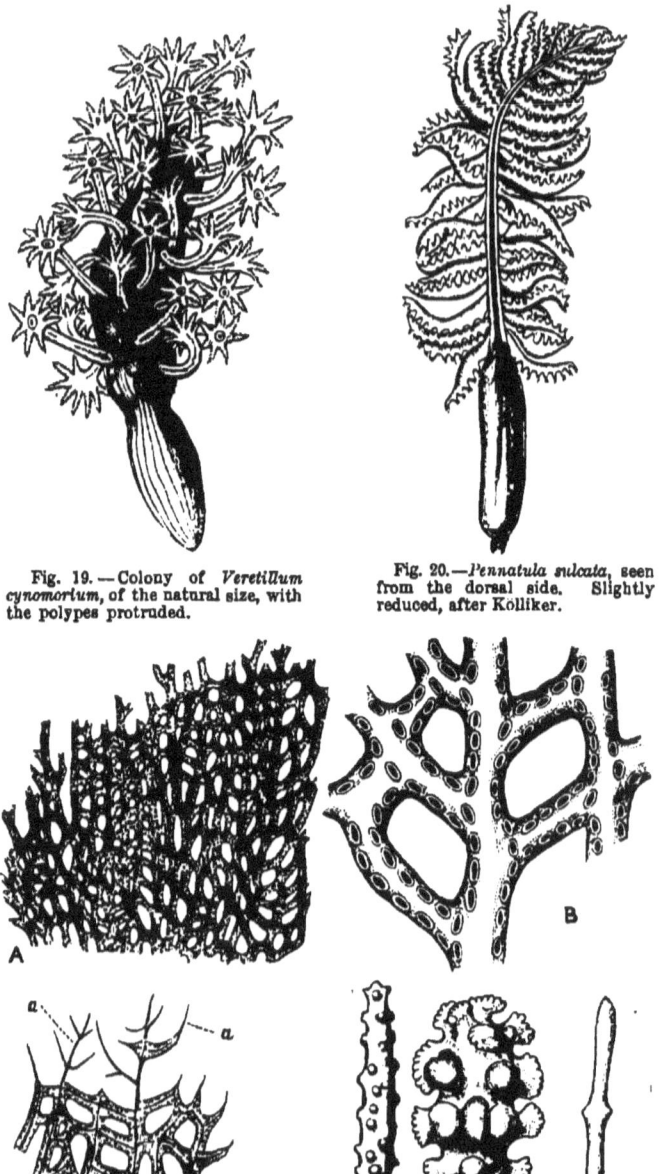

Fig. 19.—Colony of *Veretillum cynomorium*, of the natural size, with the polypes protruded.

Fig. 20.—*Pennatula sulcata*, seen from the dorsal side. Slightly reduced, after Kölliker.

Fig. 21.—A, Fragment of the common Fan-coral (*Rhipidogorgia flabellum*), reduced about one-half. B, Portion of the same enlarged, showing the polype-cells. C, Branchlet of the same partly denuded of the soft parts, and showing the horny axis (*a*). D, E, and F, Flesh-spicules ("dermosclerites") of *Gorgonidæ*, greatly enlarged: D, of *Gorgonia radula*; E, of *Sclerogorgia suberosa*; F, of *Melithæa ochracea*. (After A. Agassiz and Kölliker.)

Fam. *i*. *Chætetidæ—*Chætetes*.
Fam. *j*. *Monticuliporidæ.—*Monticulipora*.
Fam. *k*. *Auloporidæ.—*Aulopora*.

(Kölliker, *Anatomisch-systematische Beschreibung der Alcyonarien*, 1870; Kölliker, *Report on the Pennatulida*, Report of the Scientific Results of the Voyage of H.M.S. Challenger, 1881; H. N. Moseley, *Report on certain Hydroid, Alcyonarian, and Madreporarian Corals*, ibid., 1881; Nicholson, *On the Structure and Affinities of the "Tabulate Corals" of the Palæozoic Period*, 1879.)

ORDER III.—*RUGOSA.

Fam. *a*. Stauridæ.—*Stauria* (fig. 22), *Holocystis*.
Fam. *b*. Cyathaxonidæ.—*Cyathaxonia*.
Fam. *c*. Cyathophyllidæ. — *Cyathophyllum, Heliophyllum, Zaphrentis*.
Fam. *d*. Cystiphyllidæ.—*Cystiphyllum, Goniophyllum*.

(Milne-Edwards and Haime, *Polypiers fossiles des terrains paléozoiques*, 1851.)

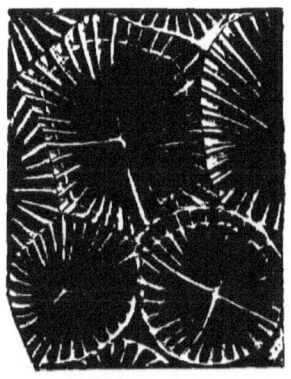

Fig. 22.—A few calices of *Stauria astræiformis*, enlarged, showing the four primary septa forming a four-branched cross. Upper Silurian. (After Milne-Edwards and Haime.)

ORDER IV.—CTENOPHORA.

Sub-ord. 1. Eurystomata.—*Beroë, Idyia*.
Sub-ord. 2. Saccatæ.—*Pleurobrachia, Hormiphora*.
Sub-ord. 3. Lobatæ.—*Bolina*.
Sub-ord. 4. Tæniatæ.—*Cestum* (Venus's Girdle, fig. 23).

(Gegenbaur, *Studien über Organisation und Systematik der Ctenophoren*, Archiv für Naturgeschichte, 1856; L. Agassiz, *Contributions to the Natural History of the United States of America*, vol. iii., 1860.)

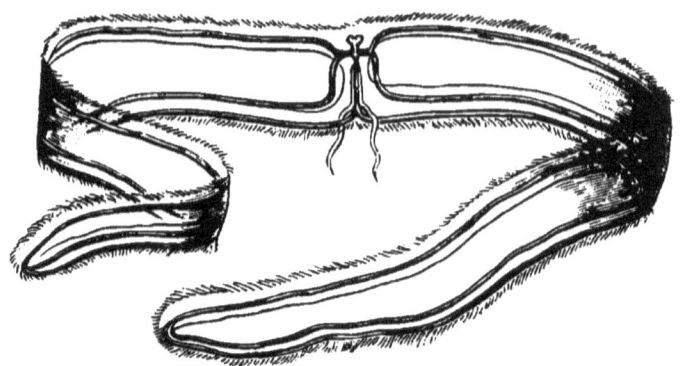

Fig. 23.—Ctenophora. *Cestum Veneris*, reduced in size.

SUB-KINGDOM (TYPE) III.—ECHINODERMATA.

SIMPLE marine organisms, which are mostly bilaterally symmetrical when young, but which in the adult condition have this bilateral symmetry more or less extensively masked by a radial (usually pentamerous) arrangement of their parts. An alimentary canal, with or without a distinct anus, separate from the proper body-cavity. A system of water-vessels, often communicating directly with the exterior, and generally connected with protrusible tubes ("feet"), is present. The nervous system is radiate, consisting of an œsophageal ring and radiating branches. The integument is characteristically hardened by the deposition in it of carbonate of lime in the form of plates, granules, or spicules.

ORDER I.—ECHINOIDEA (Sea-Urchins).
  Sub-ord. 1. Regularia (Desmosticha).
    Fam. *a.* Cidaridæ.—*Cidaris, Porocidaris.*
    Fam. *b.* Arbaciadæ.—*Arbacia, Cœlopleurus.*
    Fam. *c.* Diadematidæ. — *Diadema, Aspidodiadema, Hemicidaris.*
    Fam. *d.* Saleniadæ.—*Salenia.*
    Fam. *e.* Temnopleuridæ.—*Temnechinus.*
    Fam. *f.* Echinidæ.—*Echinus.*
    Fam. *g.* Echinothuridæ. — *Asthenosoma, Phormosoma, Echinothuria.*

  Sub-ord. 2. *Perischoechinidæ.
    Fam. *a.* *Archæocidaridæ.—*Archæocidaris.*
    Fam. *b.* *Palæchinidæ.—*Palæchinus.*

Sub-ord. 3. Irregularia.

Fam. *a*. Echinoconidæ.—*Pygaster*, *Galerites.
Fam. *b*. Clypeastridæ.—*Clypeaster, Echinocyamus, Fibularia*.
Fam. *c*. Scutellidæ.— *Mellita, Rotula, Echinarachnius,* *Scutella.
Fam. *d*. Echinoneidæ.—*Echinoneus*.
Fam. *e*. Echinobrissidæ.—*Nucleolites*.
Fam. *f*. Echinolampadæ (Cassidulidæ).—*Echinolampas, Rhynchopygus*.
Fam. *g*. *Collyritidæ (Dysastridæ).—*Collyrites*.
Fam. *h*. *Ananchytidæ.—*Ananchytes*.
Fam. *i*. Spatangidæ.—*Spatangus, Amphidetus, Brissus,* *Micraster.

(L. Agassiz, *Monographie d'Echinodermes vivans et fossiles*, 1838-42; A. Agassiz, *Revision of the Echini*, 1874; Lovén, *Études sur les Echinoïdes*, 1874; A. Agassiz, *Report on the Echinoidea*, Report of the Scientific Results of the Exploring Voyage of H.M.S. Challenger, 1881.)

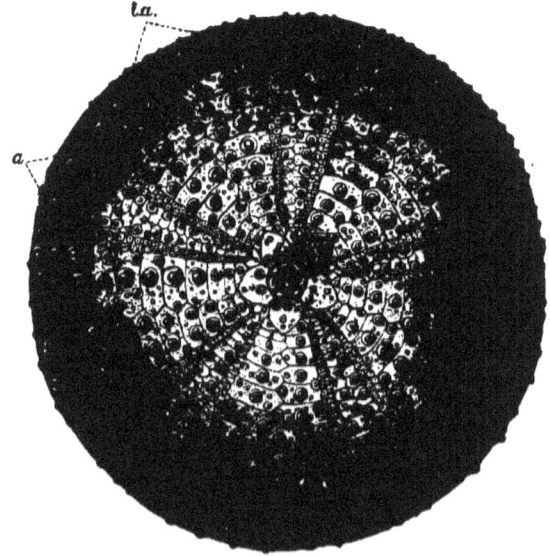

Fig. 24.—Echinoidea. Test of *Echinus esculentus*, viewed from above. *a*, One of the ambulacral areas; *ia*, One of the interambulacral areas.

ECHINODERMATA. 29

ORDER II.—ASTEROIDEA (Star-fishes).

Fam. *a.* Asteracanthiidæ.—*Uraster (Asteracanthion), Heliaster.*
Fam. *b.* Solasteridæ.—*Solaster, Cribella (Echinaster).*
Fam. *c.* Linckiadæ.—*Linckia.*
Fam. *d.* Asterinidæ.— *Asterina, Palmipes, Goniaster, Culcita.*
Fam. *e.* Astropectinidæ.—*Astropecten (Asterias), Ctenodiscus, Luidia, Archaster.*
Fam. *f.* Pterasteridæ.—*Pteraster, Hymenaster.*
Fam. *g.* Brisingidæ.—*Brisinga.*
Fam. *h.* *Palæasteridæ.—*Palæaster.*

Fig. 25.—Asteroidea. *Archaster bifrons*, viewed from the dorsal aspect. Three-fourths of the natural size. (After Sir Wyville Thomson.)

(Müller and Troschel, *System der Asteriden*, 1842; A. Agassiz, *North American Star-fishes*, Cambridge, Mass., 1877; E. Perrier, *Révision de la Collection de Stellérides du Museum d'Histoire naturelle de Paris*, Archiv de Zool. Exper., 1876.)

Order III.—Ophiuroidea (Brittle-stars).

Fam. *a*. Ophiuridæ.—*Ophiura, Ophioglypha* (fig. 26), *Ophiolepis, Ophiocoma.*

Fam. *b*. Euryalidæ.—*Asterophyton (Euryale), Asteronyx.*

(Lyman, *Ophiuridæ and Astrophytidæ*, Cat. of the Museum of Comp. Zool. at Harvard, 1865; Lütken, *Additamenta ad historiam Ophiuridarum*, 1859; Ludwig, *Morphologische Studien an Echinodermen*, 1880.)

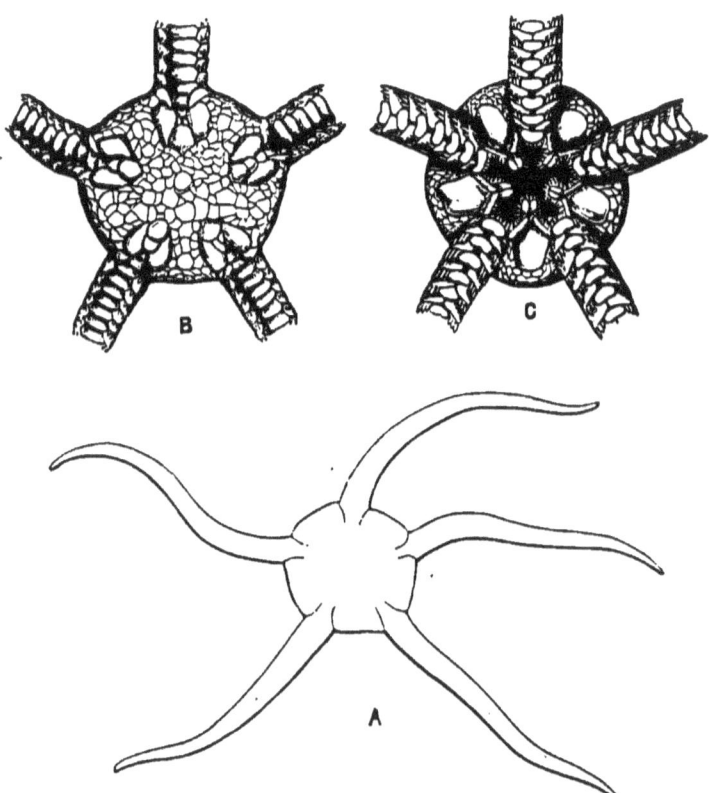

Fig. 26.—Ophiuroidea. *Ophioglypha lacertosa*: A, Outline, of the natural size; B, The disc viewed from above, twice the natural size; C, The disc viewed from below, showing the mouth and genital fissures, twice the natural size. (Original.)

ECHINODERMATA. 31

ORDER IV.—HOLOTHUROIDEA (Sea-cucumbers).

Sub-ord. 1. Apneumona.

Fam. *a*. Synaptidæ.—*Synapta, Chirodota, Anapta.*
Fam. *b*. Oncinolabidæ.—*Oncinolabes.*

Sub-ord. 2. Pneumonophora.

Fam. *a*. Molpadiidæ.—*Molpadia.*
Fam. *b*. Aspidochirotæ.—*Holothuria.*
Fam. *c*. Dendrochirotæ.—*Cucumaria (Pentacta), Psolus, Thyone.*

(Selenka, *Beiträge zur Anatomie und Systematik der Holothurien*, Zeitschr. für Wiss. Zool., 1867-68. Semper, *Reisen im Archipel der Philippinen*, 1868.)

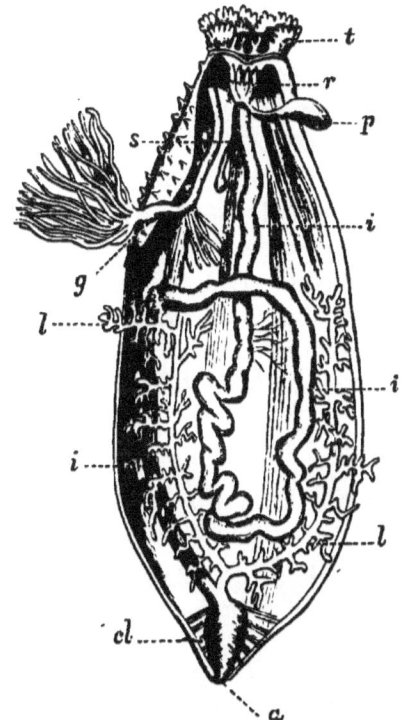

Fig. 27.—Holothuroidea. Semi-diagrammatic longitudinal section of a Holothurian. *t*, Tentacles; *r*, Calcareous ring at the base of the tentacles; *p*, Polian vesicle; *s*, Sand-canal; *i i i*, Alimentary canal; *g*, Duct of the reproductive organs; *cl*, Cloaca; *a*, Anus; *l l*, respiratory tree.

The *Holothuroidea* are often divided into primary sections, according as they possess tube-feet or not. The families which are destitute of tube-feet form the section *Apoda*, comprising the families *Synaptidæ, Oncinolabidæ,* and *Molpadiidæ*. On the other hand, the families of the *Aspidochirotæ* and *Dendrochirotæ* possess tube-feet, and form the section of the *Pediculata*.

ORDER V.—CRINOIDEA (Sea-lilies).

Sub-ord. 1. Tesselata.

Fam. *a.* *Cyathocrinidæ.—*Cyathocrinus, Zeacrinus.*
Fam. *b.* *Poteriocrinidæ.—*Poteriocrinus, Dendrocrinus.*
Fam. *c.* *Marsupitidæ.—*Marsupites.*
Fam. *d.* *Rhodocrinidæ.—*Rhodocrinus.*
Fam. *e.* *Taxocrinidæ.—*Taxocrinus.*
Fam. *f.* *Anthocrinidæ.—*Anthocrinus, Crotalocrinus.*
Fam. *g.* *Haplocrinidæ.—*Haplocrinus, Coccocrinus.*
Fam. *h.* *Pisocrinidæ.—*Pisocrinus, Triacrinus.*
Fam. *i.* *Actinocrinidæ.—*Actinocrinus, Periechocrinus.*
Fam. *j.* *Melocrinidæ.—*Melocrinus.*
Fam. *k.* *Platycrinidæ.—*Platycrinus.*
Fam. *l.* *Carpocrinidæ.—*Habrocrinus.*
Fam. *m.* *Eucalyptocrinidæ.—*Eucalyptocrinus.*
Fam. *n.* *Glyptocrinidæ.—*Glyptocrinus.*
Fam. *o.* *Gasterocomidæ.—*Gasterocoma.*
Fam. *p.* *Cupressocrinidæ.—*Cupressocrinus.*

Sub-ord. 2. Articulata.

Fam. *a.* *Encrinidæ.—*Encrinus.*
Fam. *b.* *Eugeniacrinidæ.—*Eugeniacrinus.*
Fam. *c.* Pentacrinidæ.—*Pentacrinus.*
Fam. *d.* Apiocrinidæ.—*Rhizocrinus, Bathycrinus,* *Bourgueticrinus,* *Apiocrinus.*
Fam. *e.* Holopidæ.—*Holopus,* *Cyathidium.*
Fam. *f.* *Plicatocrinidæ.—*Plicatocrinus.*
Fam. *g.* Comatulidæ. — *Antedon* (including the sub-genera *Comatula, Actinometra, Solanocrinus, Phanogenia,* &c.)

ECHINODERMATA.

The *Crinoidea* are sometimes included with the *Blastoidea* and *Cystoidea* to form a special section of *Echinodermata*, to which the name of *Pelmatozoa* is applied. If this course be followed, the *Echinoidea, Asteroidea, Ophiuroidea*, and *Holothuroidea* will constitute a second great primary division or class of Echinoderms, to which the name of *Echinozoa* may be given.

(W. B. Carpenter, *On the Structure, Physiology, and Development of Antedon rosaceus*, Phil. Trans., vol. clvi., 1876; M. Sars, *Mémoires pour servir à la connaissance des Crinoïdes vivants*, Christiania, 1868; Wyville Thomson, *Notice of New Living Crinoids belonging to the Apiocrinidæ*, Journ. Linn. Soc., 1876; P. H. Carpenter, *On the Oral and Apical Systems of Echinoderms*, Quart. Journ. Micros. Sci., vol. xviii.; Götte, *Vergleichende Entwickelungsgeschichte der Comatula Mediterranea*, Archiv für Mikros. Anat., 1876; Schultze, *Monographie der Echinodermen der Eifler Kalk*, Denkschr. der K. Akad. der Wiss., 1876.)

Fig. 28.—Crinoidea. *Comatula rosacea*, a free Crinoid, viewed from its dorsal or aboral aspect.

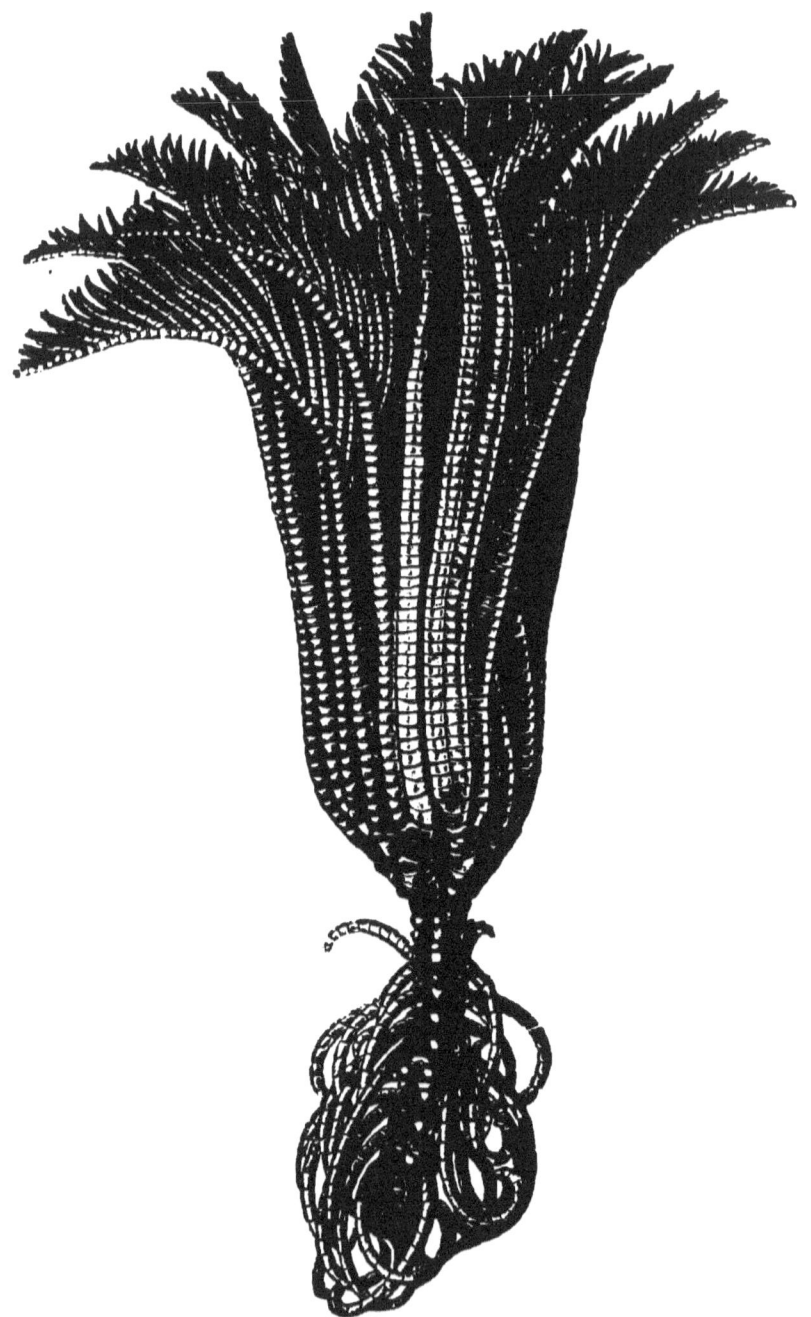

Fig. 29.—*Pentacrinus Macleayanus*, a living stalked Crinoid, slightly enlarged.

ECHINODERMATA. 35

ORDER VI.—*CYSTOIDEA.

Sub-ord. 1. Aporitidæ.—*Cryptocrinus, Malocystites.*
Sub-ord. 2. Diploporitidæ.—*Sphæronites, Glyptosphærites.*
Sub-ord. 3. Rhombiferi. — *Caryocrinus, Hemicosmites, Echinoencrinus.*

(Von Buch, *Ueber Cystideen*, Berlin, 1845; Billings, *On the Cystidea of the Lower Silurian Rocks of Canada*, 1858.)

ORDER VII.—*BLASTOIDEA.—*Pentremites, Nucleocrinus, Granatocrinus.*

(Ferd. Roemer, *Monographie der fossilen Crinoiden-familie der Blastoiden und der Gattung Pentatrematites im Besonderen*, Berlin, 1852.)

SUB-KINGDOM (TYPE) IV.—ANNULOSA.

THE body is usually more or less elongated, and is always bilaterally symmetrical, instead of being radially disposed. Typically, the body is composed of morphologically similar segments, which may be definite or indefinite, and which are arranged along a longitudinal axis. Lateral appendages may be absent or present, and when present are bilaterally disposed. A nervous system is present, consisting, in the lower forms, of one or two anteriorly-placed ganglia, but having typically the form of a ventrally-placed, double, gangliated chain.

DIVISION I.—SCOLECIDA.

CLASS I.—PLATYELMIA (Flat-worms).

ORDER I.—TÆNIOIDEA (CESTOIDEA).

Fam. *a.* Tæniada.—*Tænia* (fig. 30).
Fam. *b.* Bothriocephalidæ (Dibothridæ). — *Bothriocephalus.*
Fam. *c.* Diphyllidæ.—*Echinobothrium.*
Fam. *d.* Tetraphyllidæ.—*Phyllobothrium.*
Fam. *e.* Tetrarhynchidæ.—*Tetrarhynchus.*
Fam. *f.* Ligulidæ.—*Ligula.*
Fam. *g.* Caryophyllæidæ.—*Caryophyllæus.*

(Van Beneden, *Les vers Cestoïdes,* Mém. Acad. de Bruxelles, 1850; Leuckart, *Die Blasenbandwürmer und ihre Entwickelung,* Giessen, 1856; Spencer Cobbold, *Entozoa, An Introduction to the Study of Helminthology,* 1864.)

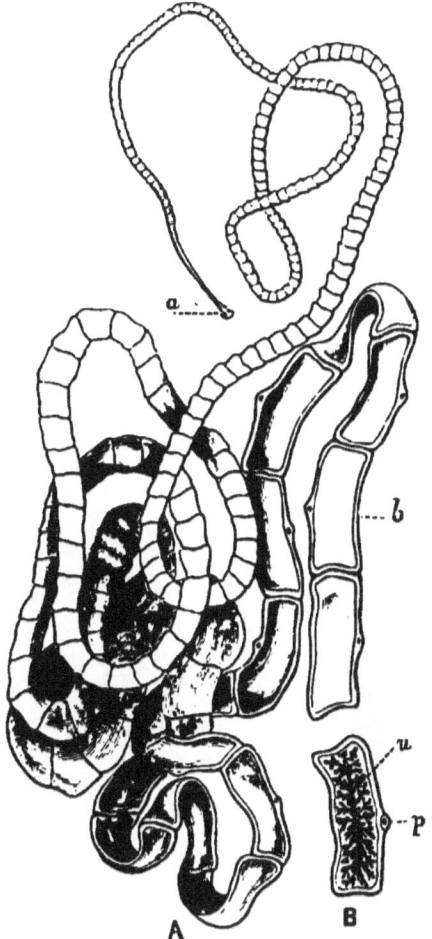

Fig. 30.—Tænioidea. A, *Tænia solium*, of the natural size: *a*, "Head" or "nurse;" *b*, One of the proglottides from the sexually mature part of the worm. B, A single mature proglottis of the same, showing the genital pore (*p*) and the branched uterus (*u*).

Order II.—Trematoda.

Sub-ord. 1. Distomata (Digenea).—*Monostomum, Diplostoma, Distoma* (fig. 31), *Gynæcophorus.*

Sub-ord. 2. Polystomata (Monogenea).—*Polystomum, Tristoma, Diplozoön, Gyrodactylus.*

Myzostoma is a singular little organism, found living as a parasite upon *Comatula* and other Crinoids, and showing many points of affinity to the

Trematode Worms. The possession, however, of a series of rudimentary feet, provided with hooks, is a character which would separate it from the Trematodes, and would rather indicate an alliance with the Chætopod Annelides.

(Pagenstecher, *Trematoden-larven und Trematoden*, Heidelberg, 1857; Van Beneden and Hesse, *Bdellodes et Trématodes marins*, Mém. Acad. de Bruxelles, 1863 and 1865; Sommer, *Anatomie des Leberegels, Distoma hepaticum*, Zeitschr. für Wiss. Zool., 1880.)

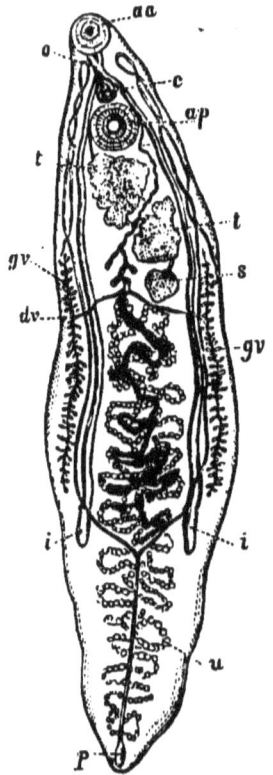

Fig. 31.—A Trematode Worm (*Distoma lanceolatum*), enlarged. *aa*, Anterior sucker, with the mouth at its bottom; *ap*, Posterior sucker; *o*, Gullet, dividing behind into the two branches of the intestine, which are unbranched, and terminate behind in blind extremities (*i i*); *p*, External opening of the water-vessels, which divide above so as to cross the blind ends of the intestine. The remaining letters refer to the different parts of the reproductive organs.

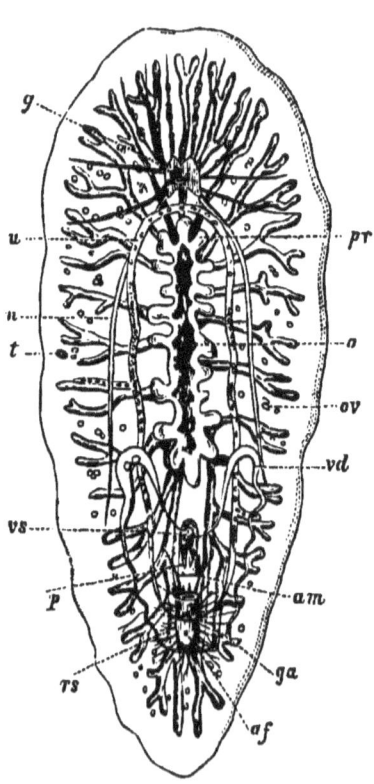

Fig. 32.—One of the Turbellarian Worms (*Leptoplana tremellaris*), enlarged. *o*, Mouth; *pr*, Proboscis; *g*, The principal nerve-ganglion, placed in the anterior part of the body, and giving off numerous radiating branches (*n*); *p*, Penis; *vd*, Vas deferens; *vs*, Vesicula seminalis; *am*, Opening of male reproductive organs; *t*, Testis; *ov*, Ovary; *u*, Uterus, partly filled with eggs; *af*, Opening of the female reproductive organs; *rs*, Receptaculum seminis; *ga*, Albuminiparous gland.

ORDER III.—TURBELLARIA.

Sub-ord. 1. Planarida.

Section A. Rhabdocœla.—*Prostomum, Opisthomum, Macrostomum, Convoluta.*

Section B. Dendrocœla. — *Planaria, Geoplana, Leptoplana* (fig. 32), *Polycelis.*

Sub-ord. 2. NEMERTIDA (Ribbon-worms).

Section A. Anopla.—*Lineus, Borlasia.*

Section B. Enopla.—*Nemertes, Tetrastemma.*

Section C. Pelagonemertida.—*Pelagonemertes* (fig. 33).

(Max Schultze, *Beiträge zur Naturgeschichte der Turbellarien*, Greifswald, 1851; Oersted, *Entwurf einer Systematischen und Speciellen Beschreibung der Plattwürmer*, Copenhagen, 1844; M'Intosh, *A Monograph of the British Nemerteans*, Ray Society, 1873-74.)

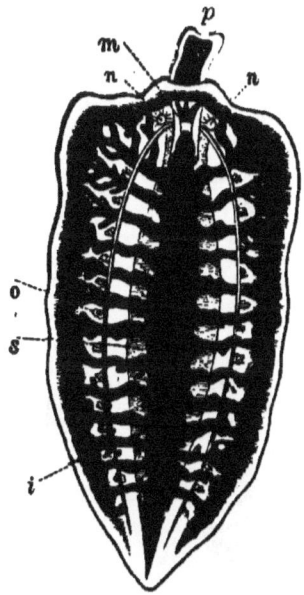

Fig. 33.—Nemertida. *Pelagonemertes Rollestoni,* a pelagic Nemertid, viewed from the ventral surface. *p*, Proboscis, partially protruded; *m*, Opening of the mouth; *i*, Alimentary canal, with its lateral diverticula, shaded darkly; *s*, The sheath of the proboscis, more lightly shaded; *n n*, The nerve-ganglia, placed one on each side of the mouth, and each giving off long lateral and backwardly-directed branch, external to which, on each side, is a row of ovaries (*o*). (After Moseley.)

CLASS II.—NEMATELMIA (Round-worms).

Nemathelminthes,

ORDER I.—ACANTHOCEPHALA.—*Echinorhynchus* (fig. 34).

(Pagenstecher, *Echinorhynchus proteus*, Zeitschr. für Wiss. Zool., 1863; Lindemann, *Anatomie der Acanthocephalen*, Moscow, 1865; Von Linstow, *Zur Anatomie und Entwickelung von Echinorhynchus angustatus*, Archiv für Naturg., 1872.)

ORDER II.—GORDIACEA (Hair-worms).

    Fam. *a.* Sphærulariidæ.—*Sphærularia.*
    Fam. *b.* Gordiidæ.—*Gordius* (fig. 35).
    Fam. *c.* Mermitidæ.—*Mermis.*

(Lubbock, *On Sphærularia bombi*, Natural History Review, 1861; A. Villot, *Monographie des Dragoneaux* (Gordiidæ), Archives de Zool. Expér., 1874.)

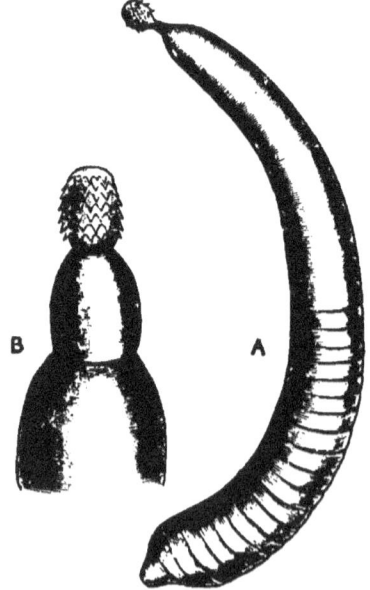
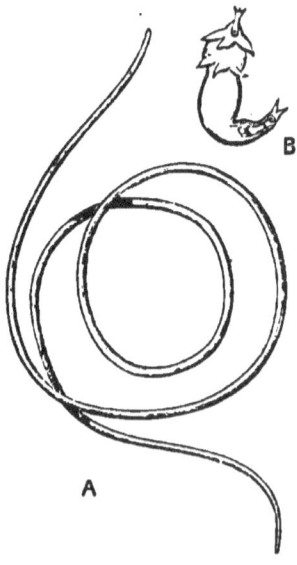

Fig. 34.—Acanthocephala. A, *Echinorhynchus gigas*, slightly enlarged. B, Head of the same, still further enlarged.

Fig. 35.—Gordiacea. A, A small individual of *Gordius aquaticus*, of the natural size. B, Larva of *Gordius subfurcatus*, with its piercing proboscis and two rows of hooks, enlarged.

ORDER III.—NEMATODA (NEMATOIDEA).

Section A. Acrophalli.

Fam. *a*. Trichocephalidæ (Trichotrachelidæ). — *Trichocephalus.*
Fam. *b*. Trichinidæ. — *Trichina.*
Fam. *c*. Strongylidæ. — *Eustrongylus, Syngamus, Dochmius, Sclerostoma.*

Section B. Hypophalli.

Fam. *a*. Spiruridæ.—*Spiroptera.*
Fam. *b*. Cucullanidæ (Cephalota).—*Cucullanus.*
Fam. *c*. Filariidæ.—*Filaria.*
Fam. *d*. Ascarida.—*Ascaris, Oxyuris.*
Fam. *e*. Cheiracanthidæ. — *Cheiracanthus.*
Fam. *f*. Anguillulidæ. — *Rhabditis* (fig. 36), *Tylenchus, Dorylaimus.*

The genera *Chætosoma* and *Rhabdogaster* include certain singular, free-living, marine worms, which have a close relationship in their internal anatomy with the ordinary Nematodes, but which have the peculiarity that the ventral surface carries a double row of bristles placed in front of the anus. If these types are included in the *Nematoda*, they must be regarded as forming a special section of the order.

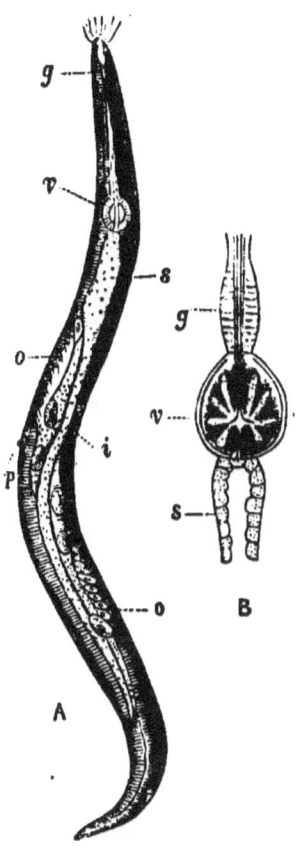

Fig. 36.—Nematoda. A, *Rhabditis bioculata*, female, enlarged. B, Portion of the alimentary tract of *Oxyuris vermicularis*, enlarged: *g*, Gullet; *v*, Muscular gizzard; *s*, Chylific stomach, or anterior end of the intestine (*i*); *oo*, Ovaries; *p*, Genital pore.

(Bastian, *Monograph of the Anguillulidæ or Free Nematoids*, Linn. Trans., 1865; Schneider, *Monographie der Nematoden*, Berlin, 1866; Leuckart, *Die Menschlichen Parasiten*, Bd. II., 1876.)

CLASS III.—ROTIFERA (ROTATORIA, Wheel-animalcules).

ORDER I.—HOLOTROCHA.—*Œcistes, Conochilus.*

ORDER II.—SCHIZOTROCHA.
    Fam. *a.* Megalotrochidæ.—*Megalotrocha.*
    Fam. *b.* Flosculariidæ. — *Floscularia, Melicerta, Stephanoceros.*
    Fam. *c.* Hydatinidæ.—*Hydatina, Eosphora* (fig. 37), *Notommata, Polyarthra, Euchlanis.*

ORDER III.—ZYGOTROCHA.
    Fam. *a.* Philodinidæ.—*Philodina, Rotifer.*
    Fam. *b.* Brachionidæ.—*Brachionus.*

ORDER IV.—GASTRODELA.
    Fam. *a.* Asplanchnidæ. — *Asplanchna.*

ORDER V.—PARASITICA.
    Fam. *a.* Albertiidæ. — *Albertia, Balatro.*

ORDER VI.—GASTROTRICHA.
    Fam. *a.* Chætonotidæ. — *Ichthydium, Chætonotus.*

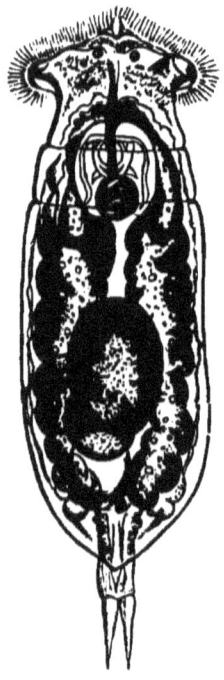

Fig. 37.—Rotifera. *Eosphora aurita*, enlarged 250 diameters. (After Gosse.)

The *Chætonotidæ*, or Hairy-backed Animalcules, constitute an aberrant group of Rotifers, and are often placed among the *Turbellaria*, or regarded as belonging to the Oligochætous Annelides.

The genera *Albertia, Seison*, and *Balatro* comprise certain abnormal Rotifers, in which there is no wheel-organ, and the cilia are either greatly reduced or wholly wanting. They are ecto- or endo-parasites.

The genus *Pedalion* comprises Rotifers with limb-like appendages, moved by special muscles, and it is sometimes regarded as the type of a special section of the Rotifers (*Arthroptera*).

The genus *Echinoderes*, lastly, includes certain minute marine organisms, in which the body is imperfectly segmented, but there are no limbs. The anterior segment of the body is furnished with hooklets, and constitutes a protrusible proboscis. The genus forms a link between the Scolecids and the higher Annulosa.

(Leydig, *Ueber den Bau und die systematische Stellung der Räderthiere*, Zeitschr. für Wiss. Zool., 1851 and 1854; Gosse, *On the Structure, Functions, and Homologies of the Manducatory Organs of the Class Rotifera*, Phil. Trans., 1856; Grenacher, *Beobachtungen über Räderthiere*, Zeitschr. für Wiss. Zool., 1869; Huxley, *Lacinularia socialis*, Trans. Micros. Soc., 1853.)

DIVISION II.—ANARTHROPODA.
CLASS I.—GEPHYREA (Spoon-worms).

ORDER I.—GEPHYREA INERMIA.
    Fam. *a*. Priapulidæ.—*Priapulus*.
    Fam. *b*. Sipunculidæ.—*Sipunculus* (fig. 38), *Phascolosoma*.

ORDER II.—GEPHYREA ARMATA.
    Fam. *a*. Echiuridæ.—*Echiurus, Thalassema*.
    Fam. *b*. Bonelliadæ.—*Bonellia*.

Fig. 38.—Gephyrea. *Sipunculus Indicus*, of the natural size. (After Keferstein.)

The genus *Sternaspis*, which has been often placed among the *Gephyrea*, is now regarded as an Annelide.

The genus *Phoronis*, on the other hand, usually placed among the Tubicolar Annelides, is sometimes looked upon as the type of a special section of *Gephyrea*.

(Keferstein, *Beiträge zur anatomischen und systematischen Kenntniss der Sipunculiden*, Zeitschr. für Wiss. Zool., 1865 and 1867; Semper, *Mittheilungen über Sipunculiden*, Zeitschr. für Wiss. Zool., 1864.)

CLASS II.—ANNELIDA (Ringed Worms).

ORDER I.—HIRUDINEA (DISCOPHORA, LEECHES).

Fam. *a.* Malacobdellidæ.—*Malacobdella.*
Fam. *b.* Acanthobdellidæ.—*Acanthobdella.*
Fam. *c.* Branchiobdellidæ.—*Branchiobdella.*
Fam. *d.* Clepsinidæ.—*Clepsine, Piscicola* (fig. 39).
Fam. *e.* Hirudinidæ.—*Sanguisuga* (*Hirudo*), *Trochetia, Nephelis* (fig. 39), *Hæmopsis, Pontobdella* (fig. 39).

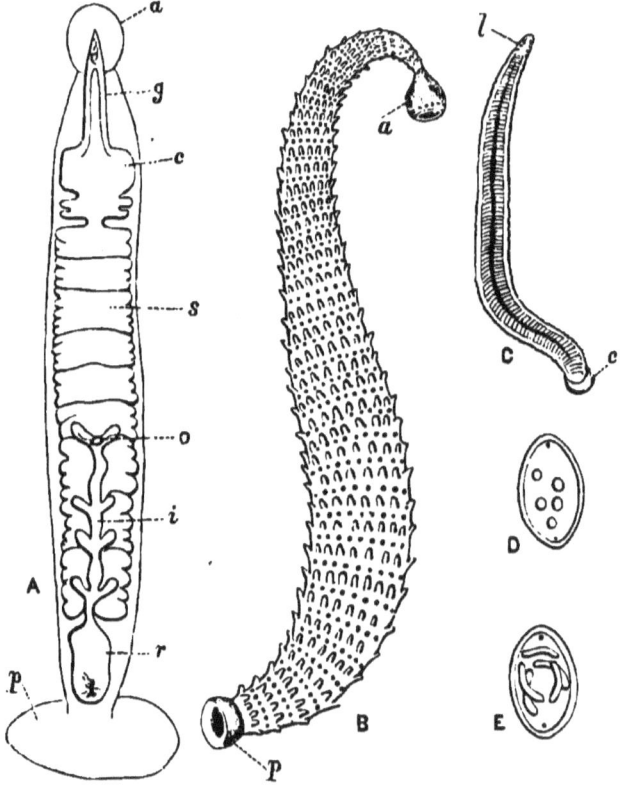

Fig. 39.—Hirudinea. A, Semi-diagrammatic view of *Piscicola geometrica*, enlarged: *a,* The anterior, and *p,* the posterior sucker; *g,* The pharynx, with the proboscis; *c,* The proventriculus; *s,* The proper stomach; *o,* Sphincter separating the stomach from the intestine; *i,* Intestine, with lateral cæca; *r,* Rectum, terminating in the aperture of the anus. B, *Pontobdella muricata,* of the natural size: *a,* Anterior, and *p,* posterior sucker. C, *Nephelis octoculata,* viewed from above, of the natural size: *l,* Upper lip, carrying the eye-spots; *c,* Posterior sucker. D, Cocoon of the preceding, with eggs, enlarged. E, An older cocoon of the same, with young leeches, enlarged. (After Leydig and Moquin-Tandon.)

The genus *Malacobdella* is sometimes regarded as belonging to the *Nemertida*. The genus *Histriobdella* is of doubtful affinities, but is usually referred to the present order.

(Moquin-Tandon, *Monographie de la Famille des Hirudinées*, 1846; Leydig, *Zur Anatomie von Piscicola geometrica*, Zeitschr. für Wiss. Zool., 1849; Dorner, *Ueber die Gattung Branchiobdella*, Zeitschr. für Wiss. Zool., 1865.)

(Claparède, *Recherches anatomiques sur les Oligochætes*, Geneva, 1862; Lankester, *On the Anatomy of the Earthworm*, Journ. Micros. Sci., 1864, 1865; Ratzel, *Beiträge zur anatomischen und systematischen Kenntniss der Oligochæten*, Zeitschr. für Wiss. Zool., 1868.)

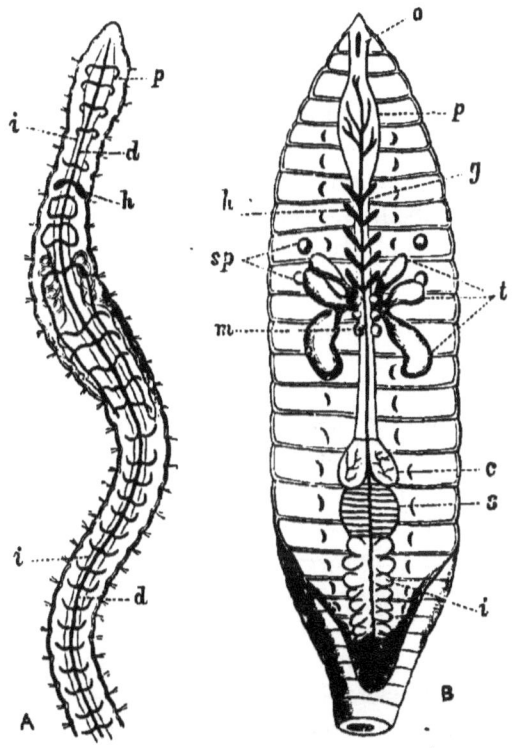

Fig. 40.—Oligochæta. A, Anterior portion of *Tubifex rivulorum*, enlarged: *p*, Pharynx; *i i*, Alimentary canal; *d d*, Dorsal vessel; *h*, One of the "hearts" or contractile dilatations of the pseudohæmal vessels. B, Anterior portion of *Lumbricus terrestris*, laid open and enlarged: *o*, Mouth; *p*, Pharynx; *g*, Gullet; *m*, Œsophageal glands; *c*, Proventriculus; *s*, Gizzard; *i*, Intestine; *h*, One of the "hearts," borne on the side of the dorsal vessel; *t*, Testes; *sp*, Spermatheca. (After Lankester.)

ORDER II.—OLIGOCHÆTA.

Fam. a. Naiididæ.—*Nais, Acolosoma.*
Fam. b. Enchytræidæ.—*Enchytræus, Chætogaster.*
Fam. c. Sænuridæ.—*Tubifex (Sænuris),* (fig. 40), *Limnodrilus.*
Fam. d. Lumbricidæ.—*Lumbricus* (Earthworm), *Criodrilus.*

ORDER III.—POLYCHÆTA.

Sub-ord. 1. Tubicola (Sedentaria).

Fam. a. Hermellidæ.—*Sabellaria (Hermella).*
Fam. b. Terebellidæ.—*Terebella, Amphitrite.*
Fam. c. Amphictenidæ.—*Pectinaria.*
Fam. d. Sabellidæ.—*Sabella, Amphicora.*
Fam. e. Serpulidæ.—*Serpula, Spirorbis, Filograna.*

Sub-ord. 2. Errantia.

Fam. a. Aphroditidæ.—*Aphrodite* (Sea-mouse).
Fam. b. Polynoidæ.—*Polynoe, Lepidonotus, Halosydna.*
Fam. c. Sigalionidæ.—*Sigalion.*
Fam. d. Nephthydidæ.—*Nephthys.*
Fam. e. Phyllodocidæ.—*Phyllodoce.*
Fam. f. Hesionidæ.—*Castalia.*
Fam. g. Syllidæ.—*Syllis, Autolytus.*
Fam. h. Nereidæ.—*Nereis, Alitta.*
Fam. i. Lumbriconereidæ.—*Lumbriconereis.*
Fam. j. Eunicidæ.—*Eunice.*
Fam. k. Amphinomidæ.—*Amphinome.*
Fam. l. Glyceridæ.—*Glycera.*
Fam. m. Telethusidæ.—*Arenicola.*
Fam. n. Spionidæ.—*Nerine, Spio.*
Fam. o. Cirratulidæ.—*Cirratulus.*
Fam. p. Tomopteridæ.—*Tomopteris.*

ANNULOSA. 47

Only the principal families of the *Annelida* are given above. The genus *Tomopteris* is often considered as forming a special section of the Polychætous Annelides, to which the name of *Gymnocopa* (Grube) has been given. The aberrant genus *Polygordius* is also sometimes referred to a special division of the *Annelida*, characterised, among other things, by the absence of setæ and parapodia.

(Quatrefages, *Histoire Naturelle des Annelés marins et d'eau douce*, Paris, 1865 ; M'Intosh, Article "*Annelides*," Encyclop. Britann., 1875 ; Ehlers, *Die Borstenwürmer*, Leipzig, 1864 and 1868.)

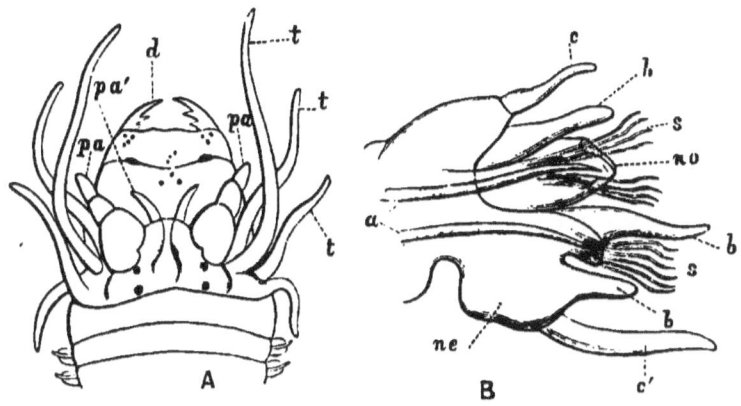

Fig. 41.—Annelida. A, Head of *Nereis incerta*, viewed from beneath, and enlarged (after Quatrefages): *d*, The principal pair of chitinous jaws (the dark dots on the lobe behind these are smaller denticles); *pa′*, Internal pair of palpi ; *pa*, External or greater pair of palpi ; *t t t*, Tentacles. B, Foot-tubercle of *Nereis*, enlarged : *no*, Notopodium ; *ne*, Neuropodium ; *c*, Dorsal cirrus ; *c′*, Ventral cirrus ; *b b b*, Branchial filaments ; *a*, Aciculæ ; *s s*, Setæ attached to the dorsal and ventral oars.

CLASS III.—CHÆTOGNATHA (Arrow-worms).

Genus *Sagitta* (fig. 42).

(Krohn, *Anatomisch-physiologische Beobachtungen über die Sagitta bipunctata*, Hamburg, 1844 ; Gegenbaur, *Ueber die Entwickelung der Sagitta*, Halle, 1856 ; Busk, *Species of Sagitta*, Quart. Journ. Micros. Sci., 1856.)

The genus *Balanoglossus* is an aberrant type, which is sometimes placed in the neighbourhood of the Nemertean worms ; while others regard it as the representative of a special section of the *Anarthropoda*, to which the name of *Enteropneusta* is applied.

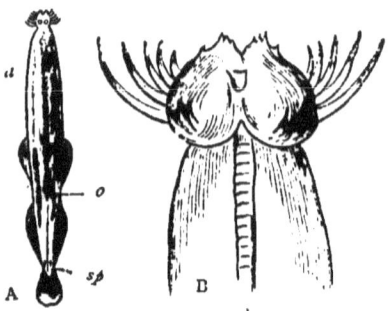

Fig. 42.—Morphology of *Chætognatha*. A, *Sagitta tricuspidata*, of the natural size: o, One of the ovaries; sp, Orifice of one of male organs of reproduction. B, Head of the same, viewed from beneath and greatly enlarged, showing the horny, setiform jaws. (After Saville Kent.)

DIVISION III.—ARTHROPODA.

CLASS I.—CRUSTACEA.

SUB-CLASS I.—EPIZOA.

ORDER I.—ICHTHYOPHTHIRA.—*Lernæa, Achtheres, Tracheliastes* (fig. 43), *Diocus* (fig. 43), *Chondracanthus, Nicothoe, Caligus.*

The *Ichthyophthira* do not form a natural division of the *Crustacea*, but may rather be more properly regarded as comprising types which are fundamentally allied to the Copepods, but which have undergone degradation in consequence of their parasitic mode of life.

(Nordmann, *Neue Beiträge zur Kenntniss parasitischen Copepoden*, Bull. de la Soc. des nat. de Moscou, 1864; Claus, *Beobachtungen über Lernæocera, Peniculus, und Lernæa*, Marburg, 1868; Claus, *Ueber den Bau und die Entwickelung von Achtheres percarum*, Zeitschr. für Wiss. Zool., 1861.)

ORDER II.—RHIZOCEPHALA.—*Sacculina, Peltogaster, Lernæodiscus.*

(Fr. Müller, *Die Rhizocephalen*, Archiv für Naturg., 1874.)

ANNULOSA. 49

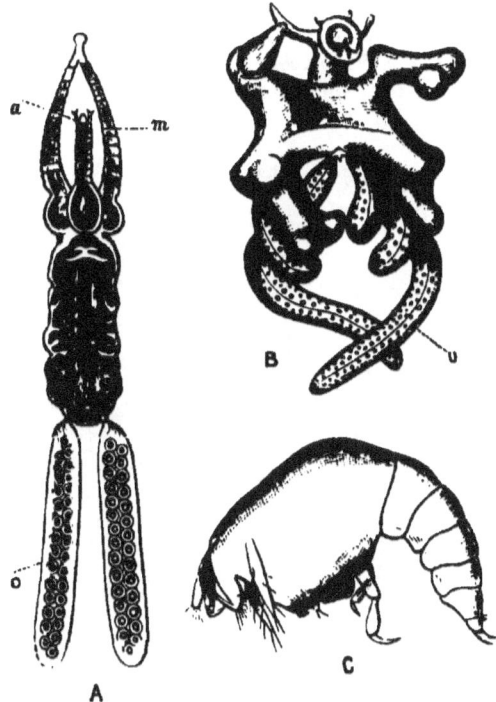

Fig. 43.—Ichthyophthira. A, Female of *Tracheliastes polycolpus*, enlarged about eight times (after Nordmann): *m*, Second pair of maxillipedes, united at their extremities to form an adhesive disc; *a*, Prehensile antennæ; *o*, Ovisacs. B, Female of *Diocus gobinus*, enlarged four times: *o*, Ovisacs. c, Pigmy male of the preceding, enlarged thirty-eight times. (After Steenstrup and Lütken.)

ORDER III.—CIRRIPEDIA.

Sub-ord. 1. Thoracica.

    Fam. *a*. Balanidæ (Acorn-shells).—*Balanus, Pyrgoma, Coronula, Chthamalus.*
    Fam. *b*. Verrucidæ.—*Verruca.*
    Fam. *c*. Lepadidæ (Barnacles).—*Lepas* (fig. 44), *Pœcilasma, Pollicipes, Scalpellum.*

Sub-ord. 2. Abdominalia.—*Cryptophialus.*

Sub-ord. 3. Apoda.—*Protcolepas.*

D

50 CLASSIFICATION OF THE ANIMAL KINGDOM.

(Darwin, *A Monograph of the Sub-class Cirripedia*, Ray Society, 1851-54 ; Pagenstecher, *Beiträge zur Anatomie und Entwickelungsgeschichte von Lepas pectinata*, Zeitschr. für Wiss. Zool., 1863 ; Von Willemoës-Suhm, *On the Development of Lepas fascicularis and the 'Archizoea' of the Cirripedia*, Phil. Trans., 1876.)

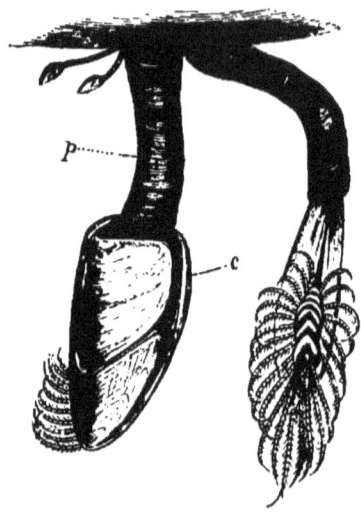

Fig. 44.—Two fully grown individuals of the common Barnacle (*Lepas anatifera*), growing upon a foreign body. *p*, The stalk of attachment; *c*, The body of the animal enclosed in a shell, from which the legs can be protruded.

SUB-CLASS II.—ENTOMOSTRACA.

ORDER I.—OSTRACODA.

Sub-ord. 1. Podocopa.

Fam. *a*. Cypridæ.—*Cypris* (fig. 45), *Candona*.
Fam. *b*. Cytheridæ.—*Cythere*, *Limnocythere*.

Sub-ord. 2. Mydocopa.

Fam. *a*. Cypridinidæ.—*Cypridina* (fig. 45), **Entomis*.
Fam. *b*. Entomoconchidæ.—*Heterodcsmus*.
Fam. *c*. Conchœciadæ.—*Halocypris*.

Sub-ord. 3. Cladocopa.

Fam. *a*. Polycopidæ.—*Polycope*.

Sub-ord. 4. Platycopa.
Fam. a. Cytherellidæ—*Cytherclla.*

(G. S. Brady, *A Monograph of the Recent British Ostracoda*, Trans. Linn. Soc., 1866; G. O. Sars, *Översigt af Norges marine Ostracoder*, 1865.)

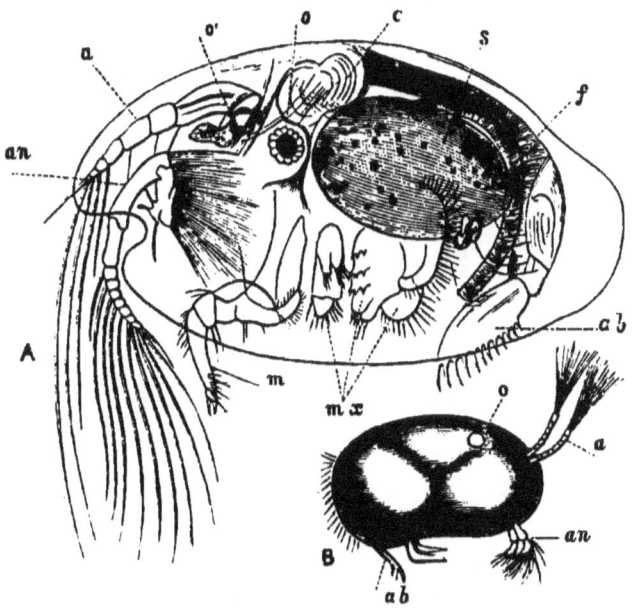

Fig. 45.—Ostracoda. A, *Cypridina Messinensis*, viewed from the side, and greatly enlarged, one-half of the shell being removed; B, *Cypris fusca*, viewed from the side, and less highly magnified, the shell-valves being retained, but slightly displaced: *a*, Antennules; *an*, Antennæ; *o*, Eye; *o'*, Ocellus; *c*, Heart; *s*, Stomach; *f*, Whip-like appendage for the retention of the brood; *ab*, Extremity of the abdomen; *m*, Mandibular appendage; *mx*, The first, second, and third maxillæ.

ORDER II.—COPEPODA.

Fam. a. Cyclopidæ.—*Cyclops* (fig. 46), *Cyclopina.*
Fam. b. Calanidæ.—*Calanus, Pontella.*
Fam. c. Notodelphyidæ.—*Notodelphys.*
Fam. d. Harpacticidæ.—*Harpacticus.*

(G. S. Brady, *A Monograph of the Free and Semi-parasitic Copepoda of the British Islands*, Ray Society, 1878-79; Claus, *Die frei-lebenden Copepoden*, Leipzig, 1863.)

52   CLASSIFICATION OF THE ANIMAL KINGDOM.

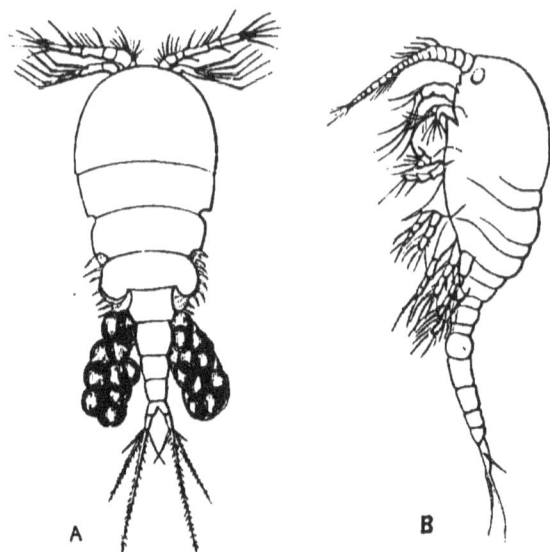

Fig. 46.—Copepoda. A, Female of *Cyclops æquoreus*, seen from above, and greatly enlarged, with the external ovisacs. B, Female of *Cyclopina littoralis*, viewed from one side, and greatly enlarged. (After G. S. Brady.)

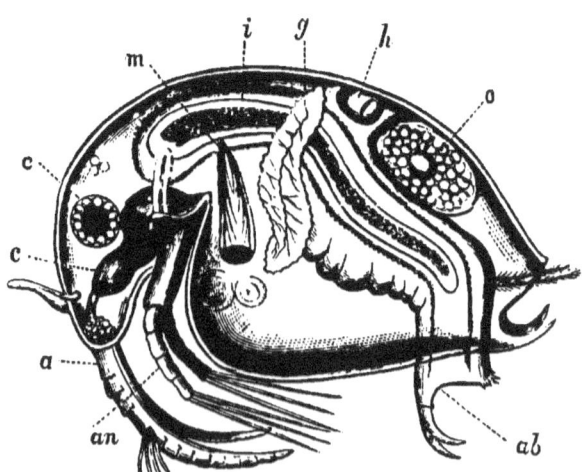

Fig. 47.—Cladocera. *Bosmina lævis*, greatly enlarged, the internal organs showing through the translucent shell: *a*, Antennules; *an*, Antennæ; *c*, Cephalic ganglion, terminating in front in a mass of ganglion-cells at the base of the antennules; *e*, Eye; *m*, Mandible; *i*, Alimentary canal; *g*, Shell-gland; *h*, Heart; *o*, Ovum contained in the brood sac; *ab*, Extremity of the abdomen, with terminal claw-like appendages. (After Leydig.)

Order III.—Cladocera.

Fam. *a.* Daphniidæ.—*Daphnia, Bosmina* (fig. 47).
Fam. *b.* Lynceidæ.—*Lynceus.*
Fam. *c.* Polyphemidæ.—*Polyphemus.*
Fam. *d.* Sididæ.—*Sida, Daphnella.*

(Leydig, *Naturgeschichte der Daphniden*, 1860; Norman and Brady, *A Monograph of the British Entomostraca belonging to the families Bosminidæ, Macrothricidæ, and Lynceidæ*, Nat. Hist. Trans. Northumberland and Durham, 1867.)

Order IV.—Phyllopoda.

Fam. *a.* Apodidæ.—*Apus* (fig. 48), *Lepidurus.*
Fam. *b.* Branchipodidæ.—*Branchipus, Artemia.*
Fam. *c.* Estheriidæ.—*Estheria, Limnadia.*
Fam. *d.* Nebaliidæ.—*Nebalia* (fig. 48).
Fam. *e.* *Peltocaridæ.—*Peltocaris, Aptychopsis.*

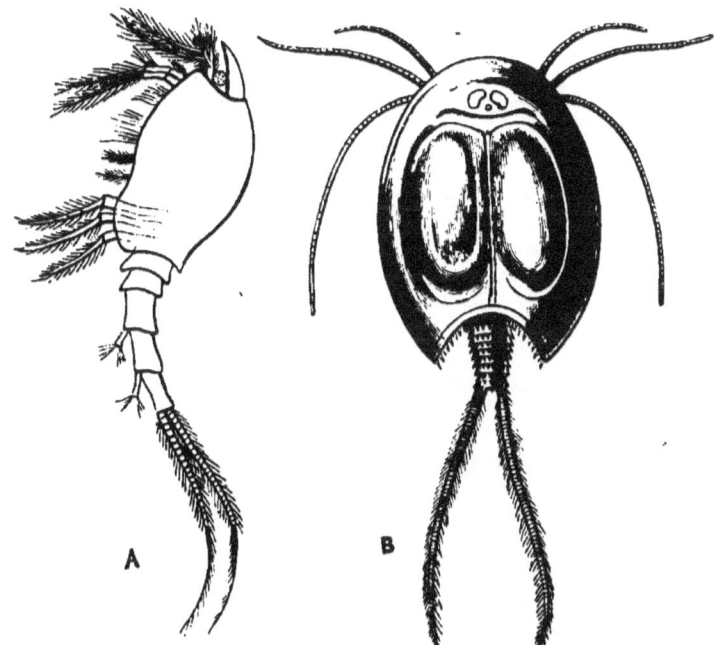

Fig. 48.—Phyllopoda. A, *Nebalia Herbstii*, enlarged about three times. B, *Apus cancriformis*, viewed from above.

The genus *Nebalia* is a transitional form, which is in many respects intermediate between the Phyllopods and the Stomapods. By its development it would appear to be referable rather to the *Malacostraca* than to the *Entomostraca*, in which case it must be placed in or near the order of the *Stomapoda*.

(Claus, *Beiträge zur Kenntniss der Entomostraken*, Marburg, 1860; Grube, *Ueber die Gattungen Estheria und Limnadia und einen neuen Apus*, Archiv für Naturg., 1865.)

ORDER V.—*Trilobita.—*Asaphus, Calymene, Illœnus* (fig. 49), *Agnostus, Paradoxides.*

(Barrande, *Système Silurien de la Bohême*, vol. i., 1852 and 1872.)

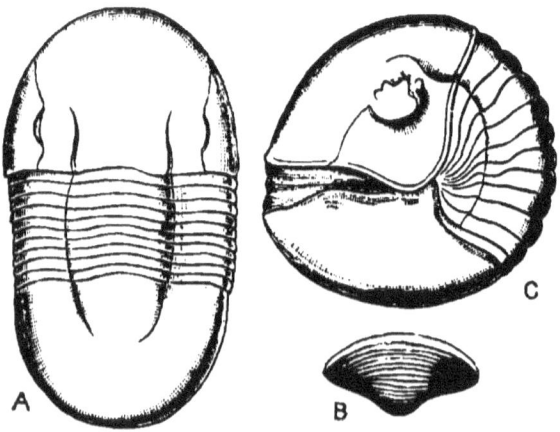

Fig. 49.—A, A complete example of *Illœnus Davisii*, in its unrolled state. B, Hypostome of the same. C, *Illœnus (Bumastus) Barriensis*, rolled up. Lower Silurian. (After Salter.)

ORDER VI.—MEROSTOMATA.

Sub-ord. 1. Xiphosura.—*Limulus.*
Sub-ord. 2. *Eurypterida.—*Pterygotus, Slimonia.*

(Owen, *On the Anatomy and Development of the American King Crab*, Trans. Linn. Soc., 1872; Packard, *The Anatomy, Histology, and Embryology of Limulus polyphemus*, Anniversary Mem. of the Boston Soc. Nat. Hist., 1880; Ray Lankester, *Limulus an Arachnid*, Quart. Journ. Micros. Science, 1881; H. Woodward, *Monograph of the Fossil Merostomata*, Palœontographical Society, 1866-72.)

Various authorities at the present day are of opinion that *Limulus* is not properly referable to the *Crustacea*, but that it is a peculiarly modified branchiate type of the *Arachnida*.

Sub-class III.—Malacostraca.

Division A.—Edriophthalmata.

Order I.—Læmodipoda.

Fam. *a.* Caprellidæ.—*Caprella* (fig. 50), *Protella.*
Fam. *b.* Cyamidæ.—*Cyamus.*

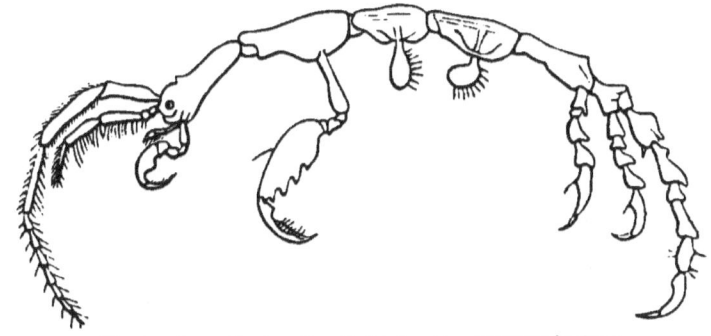

Fig. 50.—Læmodipoda. *Caprella lobata*, enlarged about six times. (After Spence Bate and Westwood.)

The *Læmodipoda* are commonly regarded as a mere section of the *Amphipoda*.

(Spence Bate and Westwood, *History of the British Sessile-eyed Crustacea*, vol. ii., 1868.)

Order II.—Amphipoda.

Section 1. Gammarinæ.
Fam. *a.* Orchestiidæ.—*Orchestia, Talitrus* (Sandhopper, fig. 51).
Fam. *b.* Gammaridæ.—*Gammarus* (Freshwater Shrimp, fig. 51), *Sulcator, Krōyera.*
Fam. *c.* Corophiidæ.—*Corophium.*
Fam. *d.* Cheluridæ.—*Chelura.*

Section 2. Hyperinæ.
Fam. *a.* Hyperiidæ.—*Hyperia.*
Fam. *b.* Phronimidæ.—*Phronima.*

56 CLASSIFICATION OF THE ANIMAL KINGDOM.

(Spence Bate and Westwood, *History of the British Sessile-eyed Crustacea*, 1868; Kröyer, *Grönlands Amphipoder beskrævne*, Kon. Danske Selsk. Naturvid. Afhandlgr., 1836.)

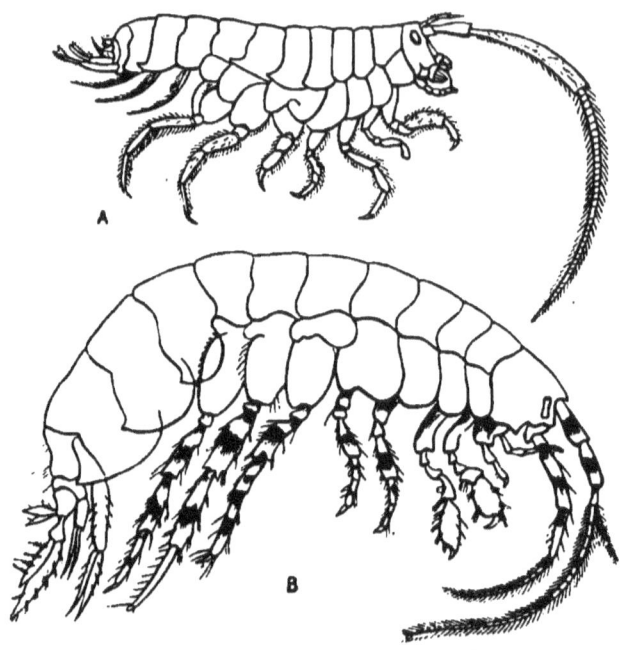

Fig. 51.—Amphipoda. A, *Talitrus locusta*, the "Sandhopper," enlarged. B, *Gammarus locusta*, enlarged about four times. (After Spence Bate and Westwood.)

ORDER III.—ISOPODA.

Section 1. Anisopoda.
    Fam. *a*. Tanaidæ.—*Tanais*.
    Fam. *b*. Anthuridæ.—*Anthura*.
    Fam. *c*. Anceidæ.—*Anceus*.

Section 2. Euisopoda.
    Fam. *a*. Cymothoidæ.—*Cymothoa, Æga, Serolis* (fig. 52).
    Fam. *b*. Sphæromidæ.—*Sphæroma*.
    Fam. *c*. Idoteidæ.—*Idotea* (fig. 52), *Arcturus* (fig. 52).
    Fam. *d*. Munnopsidæ.—*Munnopsis*.
    Fam. *e*. Asellidæ.—*Asellus, Munna, Limnoria* (Gribble).

Fam. *f.* Bopyridæ.—*Bopyrus, Cryptoniscus.*
Fam. *g.* Oniscidæ.—*Oniscus* (Wood-louse), *Ligia, Armadillo.*

(Spence Bate and Westwood, *History of the British Sessile-eyed Crustacea,* vol. ii., 1868.)

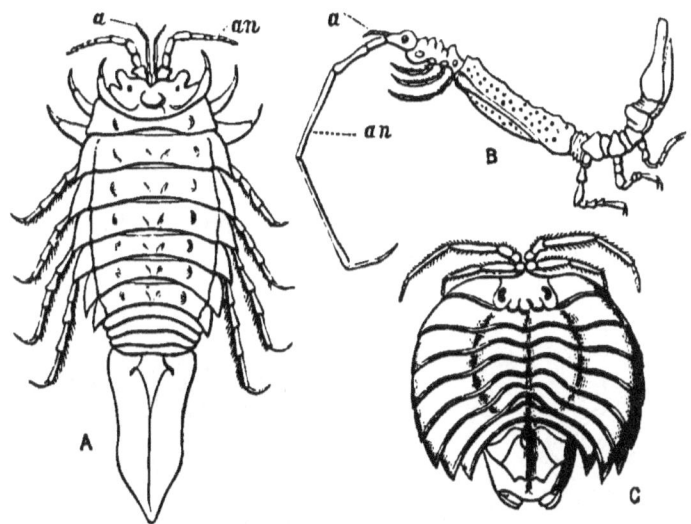

Fig. 52.—Isopoda. A, *Idotea entomon,* enlarged. B, *Arcturus longicornis,* enlarged. C, *Serolis Scythei:* a, Antennæ; *an,* Antennules. (After Gerstæcker, Spence Bate and Westwood, and Lütken.)

DIVISION B.—PODOPHTHALMATA.

ORDER I.—STOMATOPODA or STOMAPODA.

Fam. *a.* Squillidæ.—*Squilla* (Locust-shrimp, fig. 53), *Gonodactylus.*
Fam. *b.* Lophogastridæ.—*Lophogaster.*
Fam. *c.* Euphausiidæ.—*Euphausia.*
Fam. *d.* Mysidæ.—*Mysis* (Opossum-shrimp).
Fam. *e.* Leuciferidæ.—*Leucifer.*

The families *Lophogastridæ, Euphausiidæ,* and *Mysidæ* are often regarded as a separate section of the Stalk-eyed Crustaceans, to which the name of *Schizopoda* is given. The *Leuciferidæ,* also, are often looked upon as a family of the Macrurous Decapods.

ORDER II.—DECAPODA.

Tribe 1. Macrura.
Fam. a. Diastylidæ.—*Diastylis* (= *Cuma*).
Fam. b. Penæidæ.—*Penœus*.
Fam. c. Carididæ.—*Palæmon* (Prawn), *Pandalus, Hippolyte, Alpheus, Crangon* (Shrimp).
Fam. d. Astacidæ.—*Astacus* (Cray-fish), *Homarus* (Lobster), *Nephrops* (Norway Lobster).
Fam. e. Palinuridæ.—*Scyllarus, Palinurus* (Spiny Lobster), **Eryon*.
Fam. f. Thalassinidæ.—*Calianassa, Thalassina*.

The family *Diastylidæ* is often regarded as a special subdivision of the *Crustacea*, under the name of the *Cumacea*.

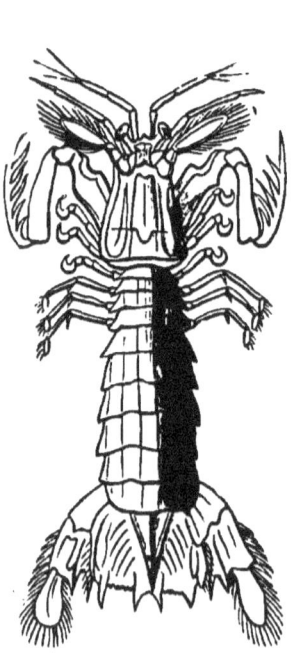

Fig. 53.—*Squilla mantis*, the Locust Shrimp.

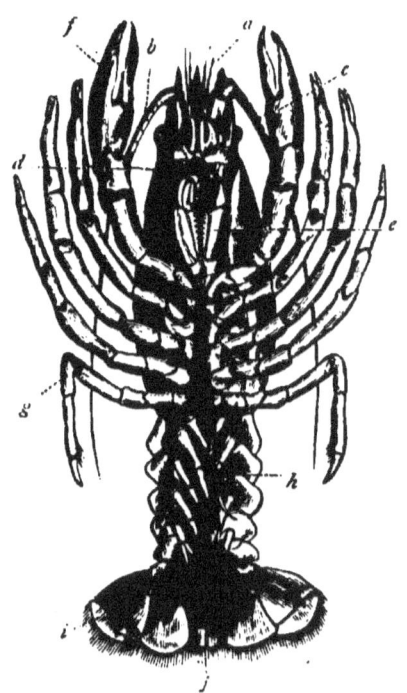

Fig. 54.—The Common Cray-fish (*Astacus fluviatilis*), viewed from below: a, Antennules; b, Large antennæ; c, Eyes; d, Opening of antennary gland; e, Last pair of foot-jaws; f, One of the great chelæ; g, Fifth thoracic limbs; h, Swimmerets; i, The last pair of swimmerets; j, The opening of the anus below the telson.

Tribe 2. Anomura.

    Fam. *a*. Galatheidæ.—*Galathea*.
    Fam. *b*. Paguridæ.—*Pagurus* (Hermit-crabs), *Cœnobita*, *Birgus*.
    Fam. *c*. Porcellanidæ.—*Porcellana*.
    Fam. *d*. Hippidæ.—*Hippa*.
    Fam. *e*. Lithodidæ.—*Lithodes* (Stone-crabs).

Tribe 3. Brachyura.

    Fam. *a*. Raninidæ.—*Ranina*.
    Fam. *b*. Leucosiadæ.—*Leucosia*.
    Fam. *c*. Calappidæ.—*Calappa*.
    Fam. *d*. Maiidæ.—*Inachus*, *Maia*, *Stenorhynchus*, *Hyas* (Spider-crabs).
    Fam. *e*. Cancridæ.—*Cancer*, *Carpilius*.
    Fam. *f*. Eriphidæ.—*Pilumnus*.
    Fam. *g*. Portunidæ.—*Portunus*, *Carcinus* (Shore-crabs).
    Fam. *h*. Corystidæ.—*Corystes*.
    Fam. *i*. Telphusidæ.—*Telphusa*.
    Fam. *j*. Pinnotheridæ.—*Pinnotheres*.
    Fam. *k*. Gonoplacidæ.—*Gonoplax*.
    Fam. *l*. Ocypodidæ.—*Ocypoda* (Sand-crabs).
    Fam. *m*. Grapsidæ.—*Grapsus*.
    Fam. *n*. Gecarcinidæ.—*Gecarcinus* (Land-crabs).

(Milne-Edwards, *Histoire Naturelle des Crustacés*, Paris, 1834-40; Dana, *Crustacea of the United States Exploring Expedition under Captain Charles Wilkes*, Philadelphia, 1852; Fritz Müller, *Für Darwin*, Leipzig, 1864 (Trans. by W. S. Dallas, "*Facts and Arguments for Darwin*," London, 1869); Leach, *Malacostraca podophthalma Britanniæ*, London, 1817-21; Bell, *History of the British Stalk-eyed Crustacea*, London, 1853.)

CLASS II.—ARACHNIDA.

ORDER I.—PANTOPODA or PODOSOMATA (Sea-spiders).

    Fam. *a*. Nymphonidæ.—*Nymphon* (fig. 55).
    Fam. *b*. Colossendeidæ.—*Colossendeus*.

60 CLASSIFICATION OF THE ANIMAL KINGDOM.

Fam. *c.* Pallenidæ.—*Pallene.*
Fam. *d.* Phoxichilidæ.—*Pycnogonum, Phoxichilus.*

(Hoek, *Report on the Pycnogonida*, Report of the Scientific Results of the Exploring Voyage of H.M.S. Challenger, vol. iii., 1881; Dohrn, *Neue Untersuchungen über Pycnogoniden*, Mitth. Zool. Stat. Neapel., i. 1879.)

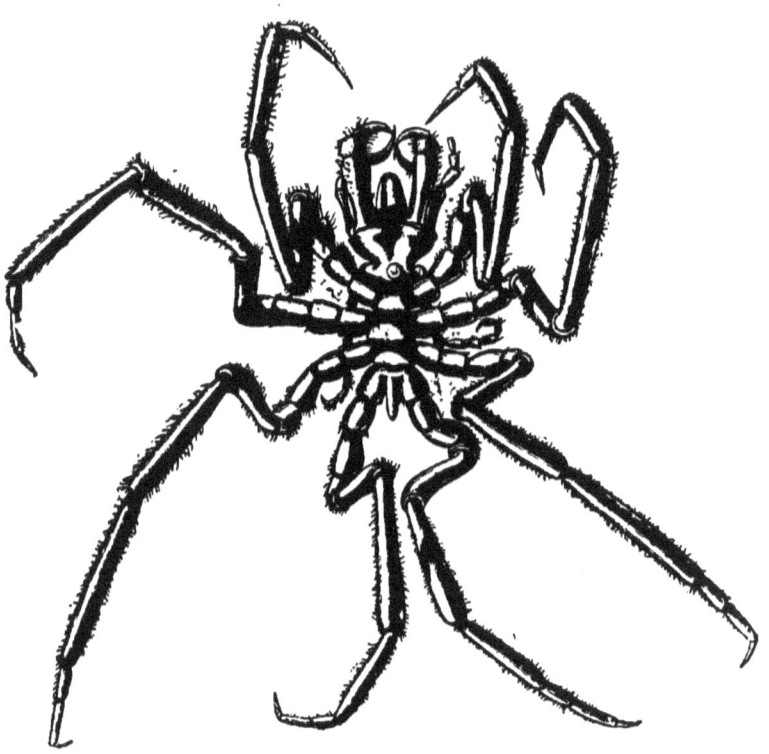

Fig. 55.—Pantopoda. *Nymphon abyssorum*, slightly enlarged.
(After Sir Wyville Thomson.)

ORDER II.—ACARINA (MONOMEROSOMATA).

Sub-ord. 1. Pentastomida (Linguatulina).
Fam. Pentastomidæ.—*Pentastoma.*

Sub-ord. 2. Tardigrada (Bear-animalcules).
Fam. Macrobiotidæ.—*Macrobiotus, Emydium.*

ANNULOSA. 61

Sub-ord. 3. Acarida.
  Fam. *a.* Dermatophilidæ.—*Demodex.*
  Fam. *b.* Sarcoptidæ.—*Sarcoptes* (Itch-mite).
  Fam. *c.* Acaridæ.—*Acarus.*
  Fam. *d.* Gamasidæ.—*Gamasus.*
  Fam. *e.* Ixodidæ.—*Ixodes* (Tick).
  Fam. *f.* Trombididæ.—*Tetranychus.*
  Fam. *g.* Hydrachnidæ.—*Limnochares,Hydrachna* (Water-mites).
  Fam. *h.* Oribatidæ.—*Oribates.*
  Fam. *i.* Bdellidæ.—*Bdella.*

(Leuckart, *Bau und Entwickelungsgeschichte der Pentastomen*, Leipzig, 1860; Doyère, *Mémoire sur les Tardigrades*, Ann. des sciences nat., 1840; Nicolet, *Histoire naturelle des Acariens*, Archives du Mus., 1855; Claparède, *Studien über die Acariden*, Leipzig, 1868; Pagenstecher, *Beiträge zur Anatomie der Milben*, Leipzig, 1860-61.)

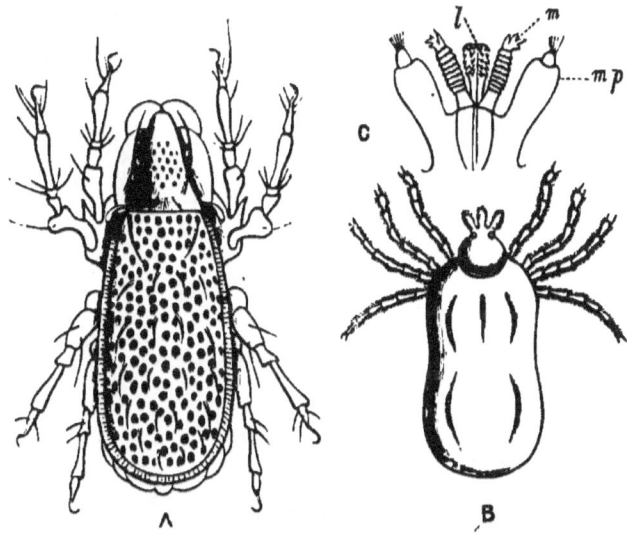

Fig. 56.—Acarina. A, *Tegeocranus elongatus*, enlarged 65 times. B, *Ixodes ricinus*, one of the Ticks, greatly enlarged. C, Mouth-organs of a Tick (*Ixodes albipictus*), enlarged; *l*, Labium; *m*, Mandibles; *mp*, Maxillary palpi. (After Michael, Packard, and Cuvier.)

ORDER III.—ADELARTHROSOMATA.
  Sub-ord. 1. Phalangidea.
    Fam. *a.* Phalangiidæ.—*Phalangium* (Harvest-men),*Opilio.*

Sub-ord. 2. Pseudoscorpionidæ.

    Fam. *a*. Cheliferidæ.—*Chelifer* (Book-scorpions, fig. 57), *Chernes*.
    Fam. *b*. Obisiidæ.—*Obisium*.

Sub-ord. 3. Solpugidea (Solifuga).

    Fam. Galeodidæ.—*Galeodes* (fig. 57).

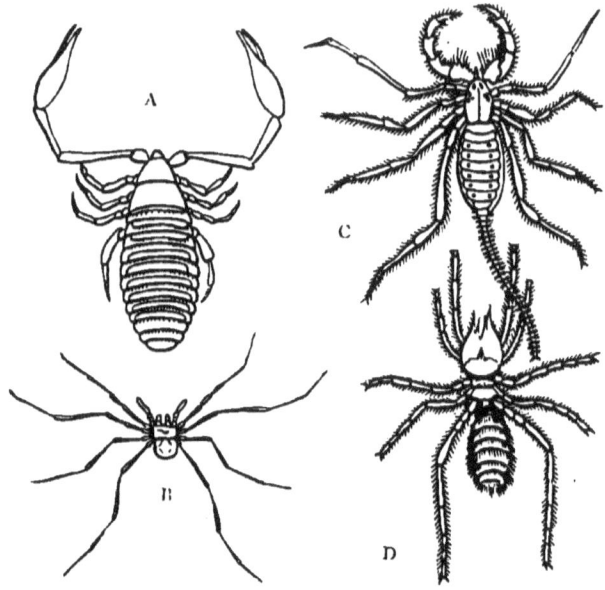

Fig. 57.—A, *Chelifer cancroides*, showing the chelate maxillary palpi, considerably enlarged. B, *Phalangium copticum*, of the natural size. C, *Thelyphonus giganteus*. D, *Galeodes araneoides*, of the natural size.

ORDER IV.—PEDIPALPI.

Sub-ord. 1. Scorpiodea (Scorpions).

    Fam. *a*. Scorpionidæ.—*Scorpio* (fig. 58).
    Fam. *b*. Androctonidæ.—*Androctonus, Buthus*.

Sub-ord. 2. Phrynidea.

    Fam. *a*. Phrynidæ.—*Phrynus*.
    Fam. *b*. Thelyphonidæ.—*Thelyphonus* (fig. 57).

(Meade, *Monograph of the British Species of Phalangium*, Ann. Nat. Hist., 1845; Dufour, *Histoire anatomique et physiologique des Scorpions*, 1856; Metschnikoff, *Embryologie des Scorpions*, Leipzig, 1870; Walckenaer and Gervais, *Histoire naturelle des Insectes aptères*, Paris, 1837-44; Packard, *Guide to the Study of Insects*, Boston, 1878.)

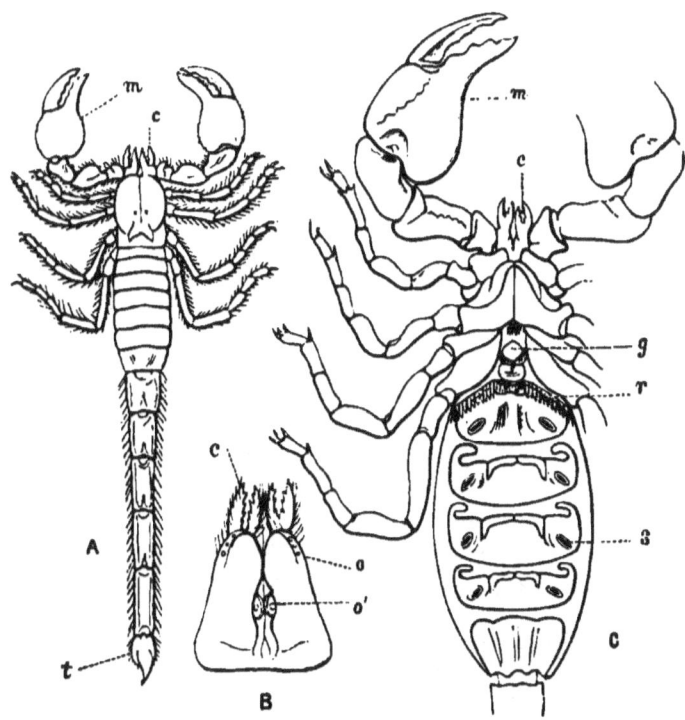

Fig. 58.—Pedipalpi. A, *Scorpio afer*, viewed from above, and somewhat reduced in size. B, Front portion of the head of the same, viewed from above, and enlarged. C, *Buthus Kochii*, with the terminal segments and the ends of the appendages on one side omitted. *m*, Maxillary palpi (behind these are the four pairs of ambulatory legs); *c*, Chelicerae; *t*, Telson; *o*, Lateral ocelli; *o'*, Central, larger ocelli; *g*, Opercular plate, covering the opening of the reproductive organs; *r*, One of the "combs;" *s*, One of the stigmatic openings. (C is after Prof. Ray Lankester.)

ORDER V.—ARANEIDA (Spiders).

Sub-ord. 1. Tetrapneumones.
Fam. Mygalidæ.—*Mygale.*

Sub-ord: 2. Dipneumones.
Section 1. Vagabunda.—*Salticus, Lycosa.*

Section 2. Sedentaria.—*Thomisus, Drassus, Tegenaria, Theridium, Epeira.*

(Cambridge, Art. "*Arachnida,*" Encyclo. Brit., 9th ed., vol. i., 1875; Blackwall, *History of the Spiders of Great Britain and Ireland,* Ray Soc., 1861-64; Staveley, *British Spiders,* 1866.)

CLASS III.—MYRIAPODA.

ORDER I.—CHILOPODA (Centipedes).

Fam. *a.* Geophilidæ.—*Geophilus.*
Fam. *b.* Lithobiidæ.—*Lithobius.*
Fam. *c.* Scolopendridæ.—*Scolopendra.*
Fam. *d.* Scutigeridæ.—*Scutigera.*
Fam. *e.* *Euphoberiidæ.—*Euphoberia.*

ORDER II.—CHILOGNATHA (Millepedes).

Fam. *a.* Glomeridæ.—*Glomeris* (Pill-millepedes).
Fam. *b.* Polyzoniidæ.—*Polyzonium.*
Fam. *c.* Polydesmidæ.—*Polydesmus.*
Fam. *d.* Polyxenidæ.—*Polyxenus.*
Fam. *e.* Iulidæ.—*Iulus* (fig. 59).
Fam. *d.* *Archiulidæ.—*Archiulus.*

(Newport, *Monograph of the Order Myriapoda, Class Chilopoda,* Linn. Trans., 1843-45; Gervais, *Études pour servir à l'histoire naturelle des Myriapodes,* Ann. des sciences nat., 1857.)

Fig. 59.—Millepede (*Iulus maximus*), a small example, of the natural size.

ORDER III.—PAUROPODA.—*Pauropus.*

(Sir John Lubbock, *On Pauropus, a New Type of Centipede,* Linn. Trans., 1868.)

ORDER IV.—ONYCHOPODA (Onychophora). — *Peripatus* (fig. 60).

ANNULOSA. 65

(Grube, *Ueber den Bau von Peripatus Edwardsii*, Archiv für Anat., 1853; Moseley, *On the Structure and Development of Peripatus Capensis*, Proc. Roy. Soc., and Ann. Nat. Hist., 1874.)

Fig. 60.—Onychopoda. *Peripatus Capensis.* (After Moseley.)

CLASS IV.—INSECTA.

SUB-CLASS I.—AMETABOLA (Aptera).

ORDER I.—ANOPLURA.

Fam. *a.* Pediculidæ (Lice).—*Pediculus, Phthirius.*

ORDER II.— MALLOPHAGA (Bird-lice). — *Trichodectes, Docophorus* (fig. 61).

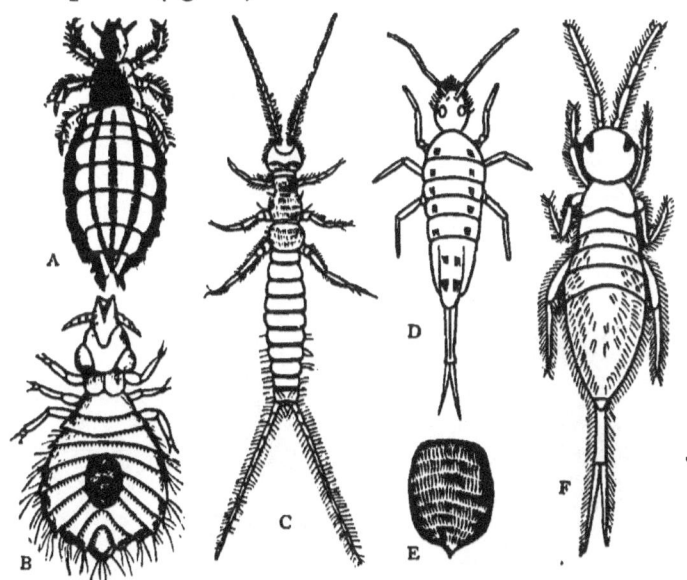

Fig. 61.—Morphology of Aptera. A, *Pediculus humanus capitis;* B, *Docophorus hamatus*, one of the Bird-lice; C, *Campodea;* D, *Degeeria*, one of the *Poduridæ;* E, Scale of a Podurid, as seen under the microscope; F, *Degeeria purpurascens*. All the figures are greatly enlarged. (After Packard and Gervais.)

E

66 CLASSIFICATION OF THE ANIMAL KINGDOM.

ORDER III.—COLLEMBOLA.—*Smynthurus, Degeeria* (fig. 61), *Podura.*

ORDER IV.—THYSANURA. — *Campodea* (fig. 61), *Lepisma, Machilis.*

SUB-CLASS II.—HEMIMETABOLA.

ORDER I.—HEMIPTERA.

Sub-ord. 1. Homoptera.

Fam. *a.* Cercopidæ.—*Aphrophora.*
Fam. *b.* Fulgoridæ.—*Fulgora* (Lantern-fly).
Fam. *c.* Cicadidæ.—*Cicada.*
Fam. *d.* Coccidæ (Scale Insects).—*Coccus, Lecanium.*
Fam. *e.* Aphididæ (Plant-lice).—*Aphis, Chermes, Phylloxera.*

Sub-ord. 2. Heteroptera.

Fam. *a.* Notonectidæ.—*Notonecta* (Boat-flies), *Corixa.*
Fam. *b.* Nepidæ.—*Nepa* (Water-scorpions).
Fam. *c.* Gerridæ.—*Gerris, Halobates.*
Fam. *d.* Hydrometridæ.—*Hydrometra.*
Fam. *e.* Reduviidæ.—*Reduvius.*
Fam. *f.* Capsidæ.—*Capsus.*

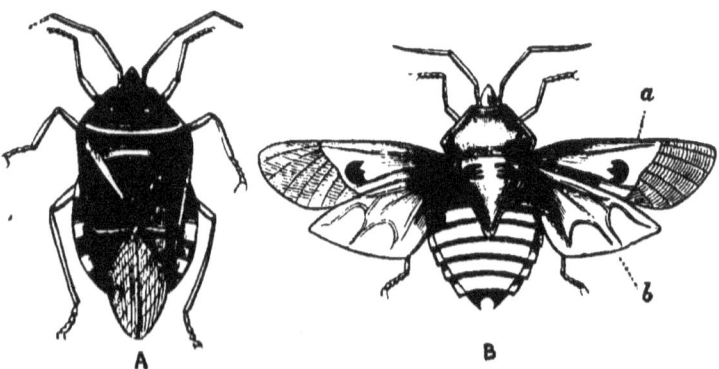

Fig. 62.—Hemiptera. A, *Pentatoma rutilans,* with the wings closed. B, *Rhaphigaster incarnatus,* with the wings expanded in flight: *a,* Anterior wing (hemelytron), with its basal portion hardened by chitine; *b,* Posterior membranous wing.

Fam. *g.* Pentatomidæ.—*Pentatoma* (Field-bugs), *Rhaphigaster* (fig. 62).

Fam. *h.* Scutelleridæ.—*Scutellera.*

Sub-ord. 3. Thysanoptera.

Fam. Thripidæ.—*Thrips.*

ORDER II.—ORTHOPTERA.

Sub-ord. 1. Cursoria.

Fam. Blattidæ (Cockroaches). — *Blatta, Periplaneta* (fig. 63).

Sub-ord. 2. Gressoria.

Fam. *a.* Mantidæ.—*Mantis.*
Fam. *b.* Phasmidæ. — *Phasma, Phyllium* (Walking-leaves).

Sub-ord. 3. Saltatoria.

Fam. *a.* Gryllidæ.—*Gryllotalpa* (Mole-cricket), *Gryllus* (Cricket).
Fam. *b.* Locustidæ.—*Locusta.*
Fam. *c.* Acrididæ.—*Œdipoda* (Migratory Locust), *Acridium* (Grasshopper).

Sub-ord. 4. Euplexoptera.

Fam. Forficulidæ.—*Forficula* (Earwig).

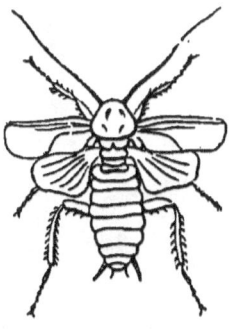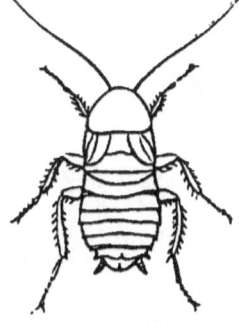

Fig. 63.—Orthoptera. The Common Cockroach (*Periplaneta orientalis*), male and female.

68 CLASSIFICATION OF THE ANIMAL KINGDOM.

ORDER III.—NEUROPTERA.

Sub-ord. 1. Corrodentia.

Fam. *a*. Psocidæ.—*Psocus*.
Fam. *b*. Embiidæ.—*Embia*.

Sub-ord. 2. Isoptera.

Fam. Termitidæ.—*Termes* (White Ants, fig. 64).

Sub-ord. 3. Amphibiotica.

Fam. *a*. Perlidæ.—*Perla* (Stone-flies).
Fam. *b*. Ephemeridæ.—*Ephemera* (May-flies), *Chloëon*.

Sub-ord. 4. Odonata (Dragon-flies).

Fam. *a*. Libellulidæ.—*Libellula*.
Fam. *b*. Æshnidæ.—*Æshna*.
Fam. *c*. Agrionidæ.—*Agrion*.

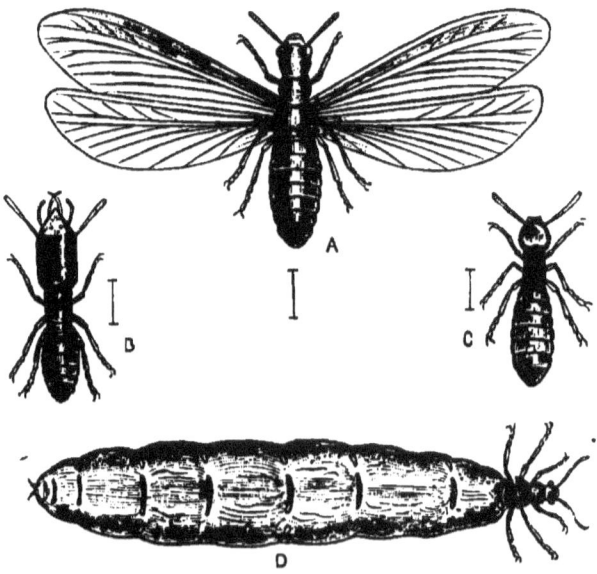

Fig. 64.—Different individuals of the colony of one of the Termites. A, The queen, before the wings are shed; D, The queen, after the wings are thrown off and the abdomen has become greatly distended with eggs; C, Worker; B, Soldier.

Sub-ord. 5. Planipennia.

    Fam. *a.* Myrmeleontidæ.—*Myrmeleo* (Ant-lion).
    Fam. *b.* Hemerobiidæ.—*Chrysopa.*
    Fam. *c.* Sialidæ.—*Corydalis.*
    Fam. *d.* Panorpidæ.—*Panorpa* (Scorpion-fly).

Sub-ord. 6. Trichoptera (Caddis-flies).—*Phryganea, Limnophilus.*

Sub-class III.—Holometabola.

Order I.—Aphaniptera.

    Fam. Pulicidæ.—*Pulex* (Flea), *Sarcopsylla* (Chigoe).

Order II.—Diptera.

Sub-ord. 1. Pupipara.

    Fam. *a.* Hippoboscidæ.—*Hippobosca* (Forest-fly), *Melophagus* (Sheep-tick).
    Fam. *b.* Nycteribiidæ.—*Nycteribia.*

Sub-ord. 2. Brachycera.

    Fam. *a.* Tabanidæ.—*Tabanus* (Gad-fly).
    Fam. *b.* Asilidæ.—*Asilus.*
    Fam. *c.* Syrphidæ.—*Syrphus, Volucella.*
    Fam. *d.* Œstridæ.—*Œstrus* (Bot-fly).
    Fam. *e.* Muscidæ.—*Musca, Stomoxys, Anthomyia.*

Sub-ord. 3. Nemocera.

    Fam. *a.* Tipulidæ.—*Tipula.*
    Fam. *b.* Cecidomyiidæ.—*Cecidomyia* (Hessian Fly).
    Fam. *c.* Chironomidæ.—*Corethra.*
    Fam. *d.* Culicidæ.—*Culex* (Gnat, fig. 65).

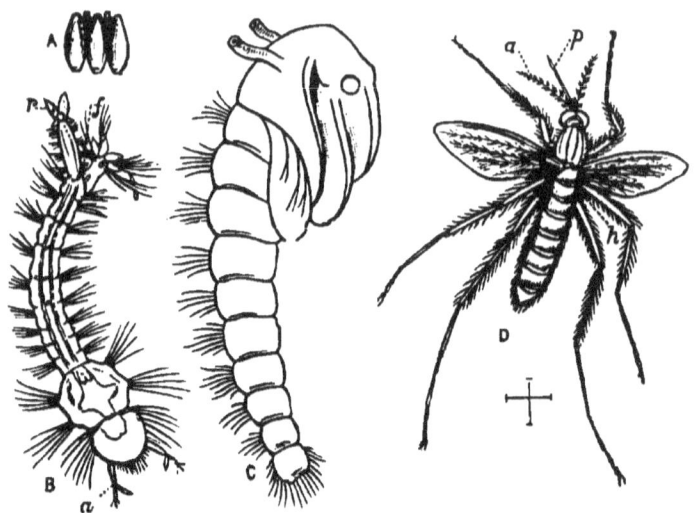

Fig. 65.—The common Gnat (*Culex pipiens*). A, A few of the eggs, attached together to form a raft, which floats on the water; B, The larva, suspended in the water head downwards, and showing the antennæ (*a*), the terminal respiratory tube (*r*), and the fins attached to the extremity of the body (*f*) ; C, The pupa, with the two respiratory tubes attached to the thorax ; D, The adult insect, with the well-developed front wings, the rudimentary hind wings or "balancers" (*h*), the antennæ (*a*), and the proboscis (*p*). All the figures are greatly enlarged.

ORDER III.—LEPIDOPTERA.

Section 1. Heterocera (Moths).

Fam. *a.* Tineidæ.—*Tinea.*
Fam. *b.* Tortricidæ.—*Tortrix.*
Fam. *c.* Geometridæ.—*Geometra.*
Fam. *d.* Noctuidæ.—*Noctua.*
Fam. *e.* Bombycidæ.—*Bombyx.*
Fam. *f.* Sphingidæ.—*Sphinx.*

Section 2. Rhopalocera (Butterflies).

Fam. *a.* Hesperiidæ.—*Hesperia.*
Fam. *b.* Lycænidæ.—*Thecla, Lycæna.*
Fam. *c.* Erycinidæ.—*Erycina.*
Fam. *d.* Nymphalidæ.—*Vanessa, Nymphalis.*
Fam. *e.* Papilionidæ.—*Colias, Papilio.*

ANNULOSA. 71

ORDER IV.—HYMENOPTERA.

Sub-ord. 1. Terebrantia.
Fam. a. Tenthredinidæ.—*Tenthredo* (Saw-fly).
Fam. b. Siricidæ.—*Sirex.*

Sub-ord. 2. Pupivora.
Fam. a. Cynipidæ (Gall-flies).—*Cynips.*
Fam. b. Chalcididæ.—*Chalcis.*
Fam. c. Ichneumonidæ.—*Ichneumon.*

Sub-ord. 3. Aculeata.
Fam. a. Formicidæ (Ants).—*Formica, Polyergus, Ponera.*
Fam. b. Vespidæ (Wasps).—*Vespa.*
Fam. c. Crabronidæ (Hornets).—*Crabro.*
Fam. d. Apidæ (Bees).—*Apis, Bombus.*
Fam. e. Andrenidæ.—*Andrena.*

Fig. 66.—Hymenoptera. a, Winged male of Ant; b, Wingless worker of Ant; c, Pupa of Ant; d, Larva of Ant, enlarged; e, The Great Saw-fly (*Sirex gigas*).

ORDER V.—STREPSIPTERA.

Fam. Stylopidæ.—*Stylops.*

ORDER VI.—COLEOPTERA (Beetles).

Sub-ord. 1. Trimera.
Fam. Coccinellidæ (Lady-birds).—*Coccinella.*

Sub-ord. 2. Tetramera.

Families: Halticidæ, Chrysomelidæ, Lamiidæ, Prionidæ (Longicorn Beetles), Curculionidæ (Weevils).

Sub-ord. 3. Heteromera.

Families: Cantharidæ, Meloidæ, Rhiphiphoridæ, Tenebrionidæ.

Sub-ord. 4. Pentamera.

Families: Telephoridæ, Elateridæ, Buprestidæ, Scarabæidæ, Lucanidæ, Silphidæ, Staphylinidæ, Hydrophilidæ, Dytiscidæ, Carabidæ, Cicindelidæ.

(Westwood, *Introduction to the Modern Classification of Insects*, 1839-40; Kirby and Spence, *Introduction to Entomology*, 1828; Burmeister, *Handbuch der Entomologie*, 1832-47; Packard, *Guide to the Study of Insects*, 6th ed., 1878.) [1]

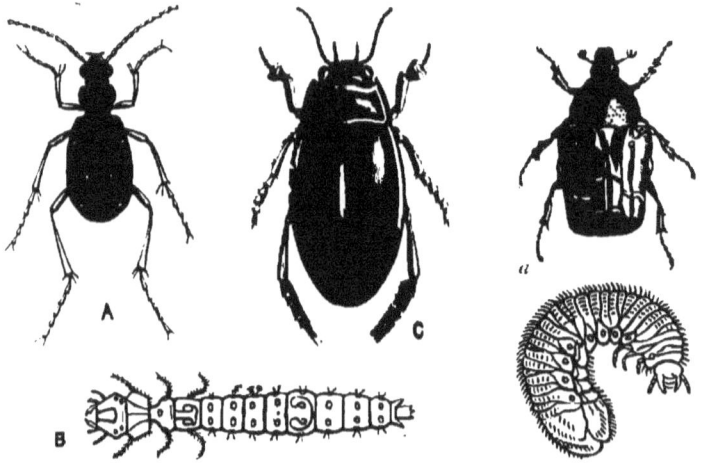

Fig. 67.—Coleoptera. A, *Cicindela campestris*, the Tiger-beetle, enlarged. B, Larva of the same, enlarged. C, *Dytiscus marginalis*, male.

Fig. 68.—a, Rose-chafer (*Cetonia aurata*) and larva.

[1] Only the more important families of the larger orders of insects are mentioned above.

SUB-KINGDOM (TYPE) V.—MOLLUSCA.

SOFT-BODIED, unsegmented animals, usually provided with an exoskeleton. Alimentary canal shut off from the body-cavity. Nervous system in the form of three principal pairs of ganglia, which are reduced to one in the lower types. A distinct heart, and specialised organs of respiration, may or may not be present. Distinct reproductive organs are present in all, though among the lower forms of the sub-kingdom the production of colonial organisms by continuous gemmation is not uncommon.

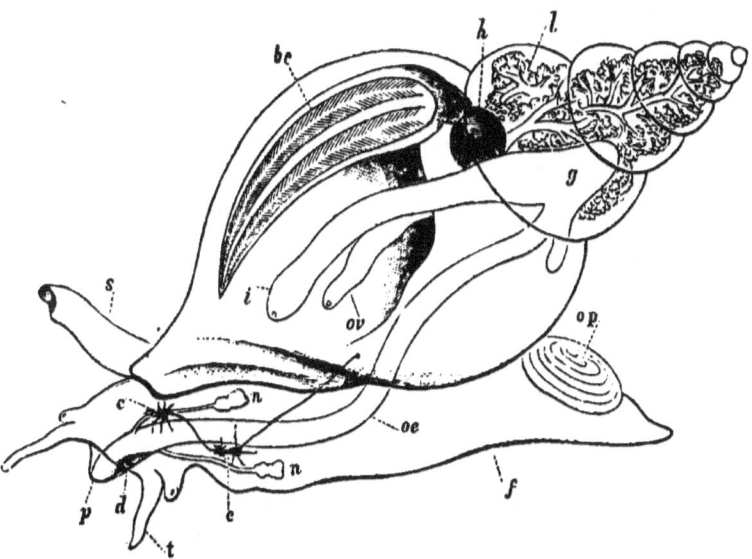

Fig. 60.—Diagram of the structure of a typical Mollusc (the Common Whelk) *f*, The muscular "foot;" *op*, The operculum; *t*, One of the tentacles, or feelers, with an eye at its base; *p*, The proboscis, retracted, with the mouth at its extremity; *œ*, Gullet; *g*, Stomach; *i*, Intestine, terminating in the anus; *n n*, Salivary glands; *l*, The liver and the ovary; *h*, The heart; *be*, The gill, contained in a hood of the mantle; *s*, Breathing-tube or siphon; *c* and *c*, The main nerve ganglia.

DIVISION A.—MOLLUSCOIDA.

CLASS I.—POLYZOA.

Sub-class I.—Ectoprocta.

Order I.—Phylactolæmata.

Fam. *a.* Plumatellidæ (Lophopea). — *Plumatella, Lophopus.*
Fam. *b.* Cristatellidæ.—*Cristatella.*

Order II.—Gymnolæmata.

Sub-ord. 1. Cheilostomata.

Fam. *a.* Catenicellidæ.—*Catenicella.*
Fam. *b.* Cellariidæ.—*Cellaria.*
Fam. *c.* Cellulariidæ.—*Cellularia.*
Fam. *d.* Scrupariidæ.—*Scruparia, Hippothoa.*
Fam. *e.* Gemellariidæ.—*Gemellaria.*
Fam. *f.* Bicellariidæ.—*Bugula.*
Fam. *g.* Flustridæ.—*Flustra* (Sea-mat, fig. 70).
Fam. *h.* Membraniporidæ.—*Membranipora.*
Fam. *i.* Celleporidæ.—*Cellepora.*
Fam. *j.* Escharidæ.—*Eschara, Lepralia.*
Fam. *k.* Reteporidæ.—*Retepora.*
Fam. *l.* Vinculariidæ.—*Vincularia.*
Fam. *m.* Selenariidæ.—*Selenaria.*

Sub-ord. 2. Cyclostomata.

Fam. *a.* Crisiidæ.—*Crisia.*
Fam. *b.* Idmoneidæ.—*Idmonea, Hornera.*
Fam. *c.* Tubuliporidæ.—*Tubulipora.*
Fam. *d.* Diastoporidæ.—*Diastopora.*
Fam. *e.* Discoporellidæ.—*Discoporella, Heteropora.*

Sub-ord. 3. Ctenostomata.

Fam. *a.* Vesiculariidæ.—*Vesicularia, Valkeria.*
Fam. *b.* Alcyonidiidæ.—*Alcyonidium.*

MOLLUSCA. 75

SUB-CLASS II.—ENTOPROCTA.

Fam. *a.* Loxosomidæ.—*Loxosoma.*
Fam. *b.* Pedicellinidæ.—*Pedicellina.*

SUB-CLASS III.—ASPIDOPHORA.

Fam. Rhabdopleuridæ.—*Rhabdopleura.*

(Allman, *A Monograph of the Fresh-water Polyzoa,* Ray Society, 1856; Busk, *Catalogue of the Marine Polyzoa in the British Museum,* 1854-76; Nitsche, *Beitrag zur Kenntniss der Bryozoen,* Zeitschr. für Wiss. Zool., 1871; Hincks, *A History of the British Marine Polyzoa,* 1880.)

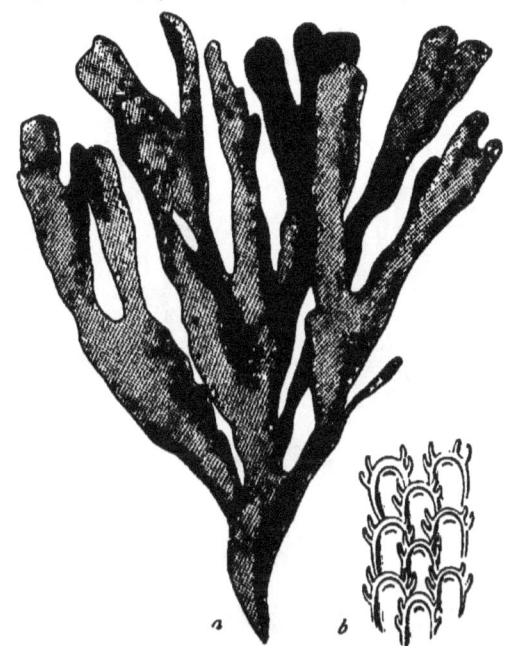

Fig. 70.—*Flustra foliacea,* one of the Sea-mats.

CLASS II.—TUNICATA (Sea-squirts).

ORDER I.—ASCIDIACEA.

Fam. *a.* Appendiculariidæ.—*Appendicularia.*
Fam. *b.* Pelonaiidæ.—*Pelonaia.*
Fam. *c.* Ascidiidæ.—*Ascidia* (fig. 71), *Ciona, Molgula, Cynthia.*
Fam. *d.* Clavellinidæ.—*Clavellina, Perophora.*

Fam. *e.* Botryllidæ. — *Botryllus, Didemnum, Amaroucium.*

Fam. *f.* Pyrosomidæ.—*Pyrosoma.*

ORDER II.—THALIACEA (BIPHORA).

Fam. *a.* Salpidæ.—*Salpa.*
Fam. *b.* Doliolidæ.—*Doliolum.*

(Milne-Edwards, *Observations sur les Ascidies composées de côtes de la Manche,* Mém. Acad. Sci. Paris, 1839; Huxley, *Upon the Anatomy and Physiology of Salpa and Pyrosoma,* Phil. Trans., 1851; Hancock, *Anatomy and Physiology of the Tunicata,* Journ. Linn. Soc., 1868; Heller, *Untersuchungen über die Tunicaten des Adriatischen und Mittelmeeres,* 1874-75; Kowalevsky, *Entwickelungsgeschichte der einfachen Ascidien,* St Petersburg, 1866.)

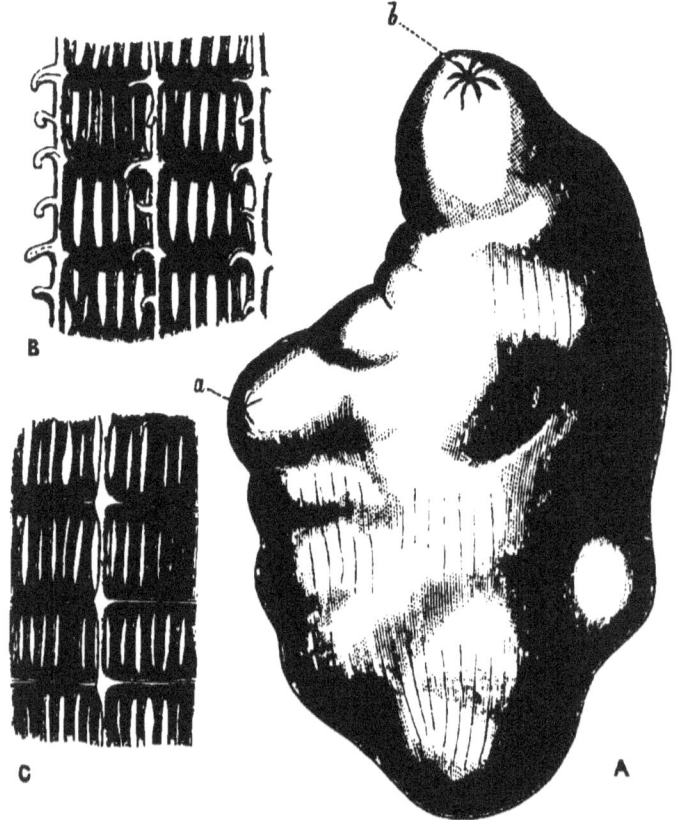

Fig. 71.—Tunicata. A, *Ascidia lata,* seen from the right side, of the natural size: *b,* Branchial aperture; *a,* Atrial aperture. B, Part of the branchial sac of the same, seen from the inside, magnified. C, Part of the branchial sac of *Ascidia virginea* (= *Ascidia sordida*), seen from the inside, magnified. (After Herdman.)

MOLLUSCA. 77

CLASS III.—BRACHIOPODA.

ORDER I.—INARTICULATA.

Fam. *a.* Lingulidæ.—*Lingula.*
Fam. *b.* Discinidæ.—*Discina.*
Fam. *c.* Craniadæ.—*Crania.*
Fam. *d.* *Trimerellidæ.—*Trimerella.*

ORDER II.—ARTICULATA.

Fam. *a.* Terebratulidæ.—*Terebratula, Argiope.*
Fam. *b.* Thecidiidæ.—*Thecidium.*
Fam. *c.* *Spiriferidæ.—*Spirifera, Athyris.*
Fam. *d.* *Koninckinidæ.—*Koninckina.*
Fam. *e.* Rhynchonellidæ.—*Rhynchonella.*
Fam. *f.* *Pentameridæ.—*Pentamerus.*
Fam. *g.* *Strophomenidæ.—*Strophomena, Orthis.*
Fam. *h.* *Productidæ.—*Producta.*

(Owen, *Anatomy of the Brachiopoda*, Trans. Zool. Soc., 1835; Hancock, *On the Organisation of the Brachiopoda*, Phil. Trans., 1858; Davidson, *Monograph of the British Fossil Brachiopoda*, Palæontographical Soc., 1851-81.)

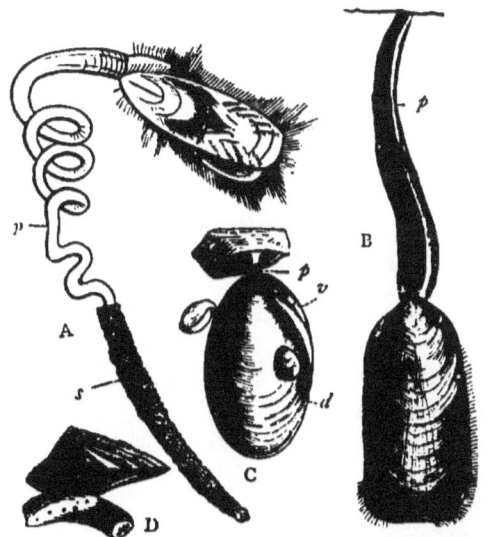

Fig. 72.—Morphology of *Brachiopoda*. A, *Lingula pyramidata* (after Morse): *p*, Peduncle; *s*, Sand-tube, encasing base of peduncle. B, *Lingula anatina* (after Cuvier): *p*, The peduncle. C, *Waldheimia cranium*, with adherent young, attached to a stone (after Davidson): *p*, Peduncle; *v*, Ventral valve; *d*, Dorsal valve. D, *Crania Ignabergensis*, attached by its ventral valve to a piece of coral (Chalk).

There are great difficulties in the way of arriving at any final conclusion as to the systematic position of the three groups of the *Polyzoa, Tunicata*, and *Brachiopoda*. Many high authorities now regard the *Polyzoa* as an aberrant group of worms, related to the true Annelides, the grounds for this conclusion being mainly derived from a study of the developmental history of the *Polyzoa*. Upon similar grounds, many naturalists consider the *Brachiopoda* as also a modified group of the worms. Lastly, the *Tunicata* are often looked upon as a degraded type of the *Vertebrata*.

DIVISION B.—MOLLUSCA PROPER.

CLASS I.—LAMELLIBRANCHIATA (CONCHIFERA).

Sub-class I.—Asiphonida.

Order I.—Ostreaceæ.

Fam. *a.* Anomiadæ.—*Anomia.*
Fam. *b.* Ostreidæ.—*Ostrea,* **Gryphæa.*
Fam. *c.* Placunidæ.—*Placuna.*
Fam. *d.* Pectinidæ.—*Pecten* (Scallop).
Fam. *e.* Limadæ.—*Lima.*
Fam. *f.* Spondylidæ.—*Spondylus.*

Order II.—Mytilaceæ.

Fam. *a.* Aviculidæ.—*Avicula* (Pearl-oyster), *Malleus,* **Inoceramus.*
Fam. *b.* Mytilidæ.—*Mytilus* (Mussel), *Dreissena.*
Fam. *c.* Pinnidæ.—*Pinna.*

Order III.—Arcaceæ.

Fam. *a.* Arcadæ.—*Arca, Pectunculus.*
Fam. *b.* Nuculidæ.—*Nucula.*
Fam. *c.* Nuculanidæ (Ledidæ).—*Nuculana (Leda).*
Fam. *d.* Trigoniadæ.—*Trigonia.*

Order IV.—Unionaceæ.

Fam. Unionidæ.—*Unio, Anodon* (Fresh-water Mussels).

Sub-class II.—Siphonida.

Order I.—*Rudistæ.

Fam. Hippuritidæ.—*Hippurites.*

Order II.—Chamaceæ.

Fam. a. Chamidæ.—*Chama,* *Diceras.*
Fam. b. Tridacnidæ.—*Tridacna, Hippopus.*

Order III.—Cardiacea.

Fam. Cardiidæ.—*Cardium* (Cockle), *Hemicardium.*

Order IV.—Lucinacea.

Fam. a. Lucinidæ.—*Lucina, Corbis.*
Fam. b. Cyprinidæ.—*Cyprina, Isocardia.*
Fam. c. Astartidæ.—*Astarte, Crassatella,* *Cardita.*

Order V.—Cycladaceæ.

Fam. Cycladidæ.—*Cyclas, Cyrena.*

Order VI.—Veneraceæ.

Fam. Veneridæ.—*Venus, Artemis.*

Order VII.—Tellinaceæ.

Fam. a. Tellinidæ.—*Tellina, Donax, Psammobia, Scrobicularia.*
Fam. b. Mactridæ.—*Mactra, Lutraria.*

Order VIII.—Myaceæ.

Fam. a. Myacidæ.—*Mya, Corbula, Saxicava.*
Fam. b. Anatinidæ.—*Anatina, Thracia, Pholadomya.*
Fam. c. Solenidæ.—*Solen* (Razor-shell), *Cultellus.*

Order IX.—Pholadaceæ.

Fam. a. Gastrochænidæ.—*Gastrochæna, Aspergillum.*
Fam. b. Pholadidæ.—*Pholas, Teredo.*

The ordinal divisions in the above list cannot be regarded as in all cases strictly natural groups, nor can they be considered as precisely equivalent to the "orders" of other classes of animals.

(Bronn, *Malacozoa Acephala*, Die Klassen und Ordnungen des Thierreichs, 1862; S. P. Woodward, *A Manual of the Mollusca*, 3d ed., 1875; Stoliczka, *The Pelecypoda of the Cretaceous Rocks of India*, Palæontologia Indica, 1875.)

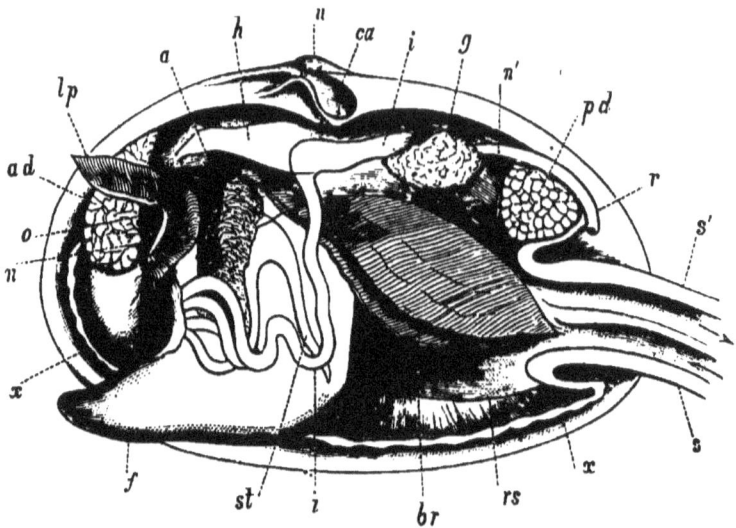

Fig. 73.—Lamellibranchiata. Diagrammatic representation of the anatomy of a siphonate Bivalve. The left valve and left mantle-lobe are removed, and the siphons are cut short. *u*, Umbo; *ca*, Cartilage-pit; *o*, Mouth; *lp*, Labial palpi; *a*, Stomach, surrounded by liver; *st*, Sac containing the crystalline stylet; *i i*, Intestine, perforating the heart (*h*); *r*, Rectum, terminating in the anus; *ad*, Anterior adductor; *pd*, Posterior adductor; *n*, Supraœsophageal or cerebral ganglion (the mouth is a little displaced upwards, so that the ganglion comes to lie below the gullet instead of above it); *n'*, Parieto-splanchnic or branchial ganglion; *f* Foot; *xx*, Cut edge of the right mantle-lobe; *rs*, Retractor muscle of the siphons; *br*, Branchiæ of the left side; *g*, Generative glands; *s*, Inhalant siphon; *s'*, Exhalant siphon.

CLASS II.—GASTEROPODA.

Sub-class I.—Branchiata.

Order I.—Scaphopoda.

Fam. Dentaliidæ.—*Dentalium*.

Order II.—Prosobranchiata.

Section A. Siphonostomata.

Fam. *a*. Strombidæ.—*Strombus, Pteroceras.*
Fam. *b*. Muricidæ.—*Murex, Fusus.*

MOLLUSCA. 81

Fam. *c.* Buccinidæ.—*Buccinum* (Whelk), *Nassa, Purpura, Oliva.*
Fam. *d.* Conidæ.—*Conus, Pleurotoma.*
Fam. *e.* Volutidæ.—*Voluta, Mitra.*
Fam. *f.* Cypræidæ (Cowries).—*Cyprœa, Ovulum.*

Section B. Holostomata.

Fam. *a.* Naticidæ.—*Natica, Sigaretus.*
Fam. *b.* Pyramidellidæ.—*Pyramidella, Chemnitzia.*
Fam. *c.* Cerithiadæ.—*Cerithium, Aporrhais.*
Fam. *d.* Melaniadæ.—*Melania.*
Fam. *e.* Turritellidæ.—*Turritella, Scalaria, Vermetus.*
Fam. *f.* Littorinidæ.—*Littorina* (Periwinkle), *Solarium.*
Fam. *g.* Paludindiæ.—*Paludina, Ampullaria.*
Fam. *h.* Neritidæ.—*Nerita.*
Fam. *i.* Turbinidæ.—*Turbo, Trochus.*
Fam. *j.* Haliotidæ. — *Haliotis* (Ear - shell), *Pleurotomaria.*
Fam. *k.* Fissurellidæ.—*Fissurella* (Keyhole Limpet).
Fam. *l.* Calyptræidæ.—*Calyptrœa* (Cup-and-saucer Limpet), *Capulus* (Bonnet Limpet).

Section C. Cyclobranchiata.

Fam. Patellidæ (Limpets).—*Patella, Acmœa.*

Section D. Polyplacophora.

Fam. Chitonidæ.—*Chiton.*

ORDER III.—OPISTHOBRANCHIATA.

Section A. Nudibranchiata.

Families:—Doridæ (Sea-lemons), Æolidæ, Dendronotidæ, Tethydidæ.—*Doris, Æolis, Doto, Tethys.*

Section B. Tectibranchiata.

Fam. *a.* Tornatellidæ.—*Tornatella.*
Fam. *b.* Bullidæ.—*Bulla* (Bubble-shells).
Fam. *c.* Aplysiadæ.—*Aplysia* (Sea-hare).
Fam. *d.* Pleurobranchidæ.—*Umbrella.*
Fam. *e.* Runcinidæ.—*Runcina.*

F

ORDER IV.—HETEROPODA (NUCLEOBRANCHIATA).

Fam. *a.* Firolidæ.—*Firola, Carinaria.*
Fam. *b.* Atlantidæ.—*Atlanta.*

SUB-CLASS II.—PULMONATA.

ORDER I.—INOPERCULATA.

Fam. *a.* Helicidæ.—*Helix* (Land-snail), *Bulimus, Pupa.*
Fam. *b.* Limacidæ.—*Limax* (Slug).
Fam. *c.* Limnæidæ.—*Limnæa, Planorbis.*
Fam. *d.* Auriculidæ.—*Auricula.*

ORDER II.—OPERCULATA.

Fam. *a.* Cyclostomidæ.—*Cyclostoma.*
Fam. *b.* Aciculidæ.—*Acicula.*
Fam. *c.* Helicinidæ.—*Helicina.*

(H. and A. Adams, *The Genera of Recent Mollusca*, London, 1858; Keferstein, *Malacozoa Cephalophora*, in Bronn's Klassen und Ordnungen des Thierreichs, 1862-66; S. P. Woodward, *Manual of the Mollusca*, 3d ed., 1875; Huxley, *On the Morphology of the Cephalous Mollusca*, Phil. Trans., 1853; Troschel, *Das Gebiss der Schnecken*, 1856; Stoliczka, *Cretaceous Gasteropoda of Southern India*, Palæontologia Indica, 1868.)

Fig. 74.—Gasteropoda. The Garden Snail (*Helix aspersa*).

MOLLUSCA.

CLASS III.—PTEROPODA.

ORDER I.—GYMNOSOMATA.

Fam. Cliidæ.—*Clio, Pneumodermon.*

ORDER II.—THECOSOMATA.

Fam. *a.* Hyaleidæ.—*Hyalea, Cleodora.*
Fam. *b.* *Hyolithidæ.—*Hyolithes (Theca).*
Fam. *c.* Cymbuliidæ.—*Cymbulia.*
Fam. *d.* Limacinidæ.—*Limacina, Spirialis.*

(Rang et Souleyet, *Histoire naturelle des Mollusques Ptéropodes,* Paris, 1852; Gegenbaur, *Untersuchungen über die Pteropoden und Heteropoden,* Leipzig, 1853; Krohn, *Beiträge zur Entwickelungsgeschichte der Pteropoden und Heteropoden,* Leipzig, 1860; Barrande, *Ptéropodes,* in the 'Systême Silurien du Centre de la Bohême,' 1867.)

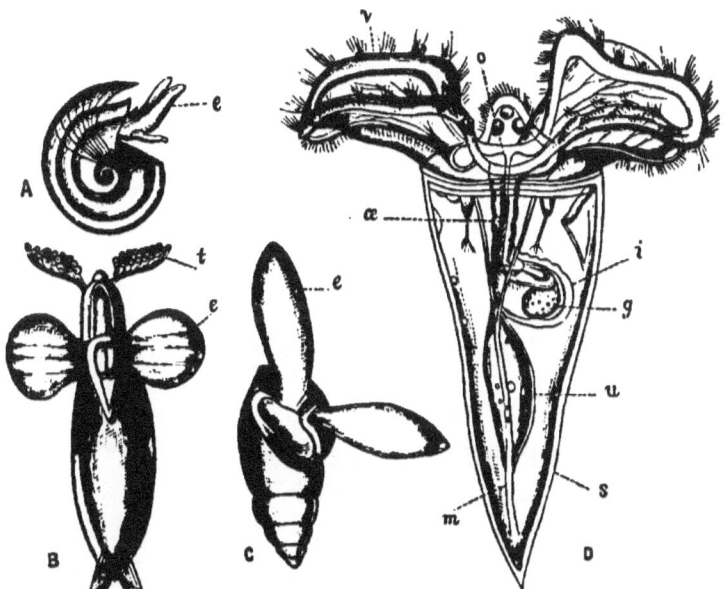

Fig. 75.—Pteropoda. A, *Spirialis rostralis* B, *Pneumodermon violaceum.* C, *Heterofusus buliminoides.* All enlarged. *e,* Epipodia or fins; *t,* Tentacles. D, Larva of *Cleodora lanceolata,* greatly enlarged (after Fol): *v,* Velum; *o,* Mouth; *œ,* Gullet; *g,* Stomach; *i,* Intestine; *m,* Columellar muscle; *s,* Shell; *u,* Yolk-sac.

CLASS IV.—CEPHALOPODA.

ORDER I.—DIBRANCHIATA (Cuttle-fishes).

Sub-ord. 1. Octopoda.

Fam. *a.* Octopodidæ.—*Octopus* (Poulpe), *Eledone.*
Fam. *b.* Argonautidæ.—*Argonauta* (Paper Nautilus).

Sub-ord. 2. Decapoda.

Fam. *a.* Spirulidæ.—*Spirula* (Post-horn).
Fam. *b.* *Belemnitidæ.—*Belemnites, Belemnitella.*
Fam. *c.* Sepiadæ.—*Sepia.*
Fam. *d.* Sepiolidæ.—*Sepiola, Rossia.*
Fam. *e.* Loliginidæ (Teuthidæ).—*Loligo* (Calamary).
Fam. *f.* Chiroteuthidæ.—*Chiroteuthis, Ommastrephes.*
Fam. *g.* Loligopsidæ.—*Loligopsis.*
Fam. *h.* Cranchiidæ.—*Cranchia.*

ORDER II.—TETRABRANCHIATA.

Sub-ord. 1. Nautiloidea.

Fam. *a.* Nautilidæ.—*Nautilus* (Pearly Nautilus), *Lituites.*
Fam. *b.* *Orthoceratidæ.—*Orthoceras, Cyrtoceras.*

Sub-ord. 2. *Ammonitoidea.

Fam. *a.* *Goniatitidæ.—*Goniatites, Bactrites.*
Fam. *b.* *Ceratitidæ.—*Ceratites.*
Fam. *c.* *Ammonitidæ.—*Ammonites, Baculites, Scaphites.*

(Ferussac and D'Orbigny, *Histoire naturelle des Céphalopodes acétabulifères vivants et fossiles,* Paris, 1835-48 ; Owen, Art. " *Cephalopoda,*" in Todd's Cyclopædia of Anat. and Phys., 1836 ; Owen, *Memoir on the Pearly Nautilus,* 1832 ; Hyatt, *Embryology of the Tetrabranchiates,* Bull. Mus. Comp.

MOLLUSCA. 85

Zool., 1872; Quenstedt, *Die Cephalopoden*, 1846; Barrande, *Céphalopodes*, in 'Système Silurien du Centre de la Bohême,' 1872-77.)

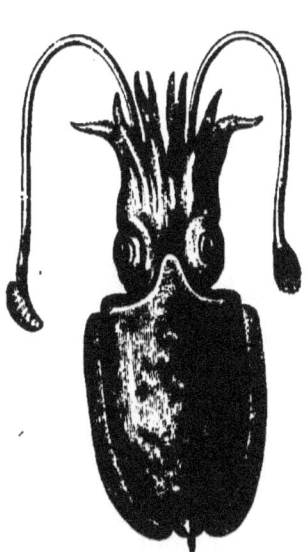

Fig. 76.—Cephalopoda. *Sepia elegans*.

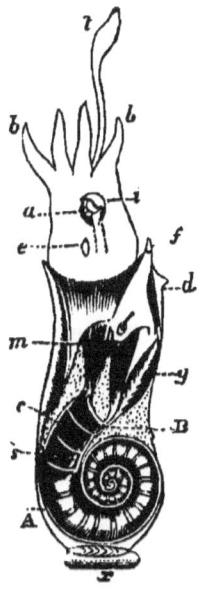

Fig. 77.—Anatomy of *Spirula australis* (after Owen), showing the position of the skeleton.

SUB-KINGDOM (TYPE) VI.—VERTEBRATA (CHORDATA).

THE body in the Vertebrata is usually composed of a number of more or less definite segments (very obscure in the Cyclostomatous Fishes), which are arranged along a longitudinal axis. The nervous system is in its main masses dorsal, and the neural and hæmal regions of the body are always completely separated from one another. The cerebro - spinal nervous axis is underlaid by the structure known as the "notochord," which, in adult life, is usually more or less completely replaced by the bony axis known as the "spine" or "vertebral column." The limbs are sometimes absent; but, when present, they are never more than four in number, and are always turned away from the neural aspect of the body.

DIVISION A.—ICHTHYOPSIDA.

CLASS I.—PISCES (FISHES).

ORDER I.—PHARYNGOBRANCHII.

Fam. Cirrostomi.—Amphioxus (Lancelet, fig. 78).

(J. Müller, *Ueber den Bau und die Lebenserscheinungen der Branchiostoma lubricum*, Abhandl. der Berl. Akad., 1842 ; Kowalevsky, *Entwickelungsgeschichte von Amphioxus lanceolatus*, St Petersburg, 1867.)

Fig. 78.—Pharyngobranchii. The Lancelet (*Amphioxus lanceolatus*), enlarged.

VERTEBRATA (CHORDATA). 87

ORDER II.—MARSIPOBRANCHII (CYCLOSTOMATA).

Fam. *a.* Myxinidæ.—*Myxine* (Hag-fish).
Fam. *b.* Petromyzonidæ.—*Petromyzon* (Lamprey).

(J. Müller, *Vergleichende Anatomie der Myxinoiden*, Abhandl. der Berlin Akad., 1835-1845.)

ORDER III.—TELEOSTEI (Bony Fishes).

Sub-ord. 1. Malacopteri (Physostomi).

Section A. Apoda.

Families: Murænidæ, Symbranchidæ, Gymnotidæ.
—*Murœna* (Eel), *Symbranchus, Gymnotus* (Electric-eel).

Section B. Abdominalia.

Families: Clupeidæ, Esocidæ, Cyprinidæ, Salmonidæ, Siluridæ.—*Clupea* (Herring), *Esox* (Pike), *Cyprinus* (Carp), *Salmo* (Salmon), *Silurus* (Sheatfish).

Sub-ord. 2. Anacanthini.

Section A. Gadoidei.

Families: Gadidæ, Ophidiidæ, Macruridæ.—*Gadus* (Cod), *Ammodytes* (Sand-eel), *Bathygadus*.

Section B. Pleuronectoidea.

Fam. Pleuronectidæ (Flat-fishes). — *Pleuronectes* (Plaice), *Rhombus* (Turbot), *Solea* (Sole).

Sub-ord. 3. Acanthopteri.

Section A. Pharyngognathi.

Families: Pomacentridæ, Labridæ, Chromidæ.—*Pomacentrus, Labrus* (Wrasse), *Chromis*.

Section B. Acanthopteri veri.

Families: Percidæ, Mugilidæ, Scomberidæ, Sclerogenidæ, Gobiidæ, Blenniidæ, Lophiidæ. — *Perca* (Perch), *Mugil* (Mullet), *Scomber* (Mackerel), *Cottus* (Bull-head), *Gobius* (Goby), *Anarrhicas* (Wolffish), *Lophius* (Angler).

88 CLASSIFICATION OF THE ANIMAL KINGDOM.

Sub-ord. 4. Plectognathi.

Section A. Sclerodermi.

Families: Balistidæ, Ostraciontidæ.—*Balistes* (File-fish), *Ostracion* (Trunk-fish).

Section B. Gymnodontes.

Families: Tetrodontidæ, Molidæ.—*Diodon* (Globe-fish), *Orthagoriscus* (Sun-fish).

Sub-ord. 5. Lophobranchii.

Families: Solenostomidæ, Syngnathidæ, Hippocampidæ.—*Solenostoma*, *Syngnathus* (Pipe-fish), *Hippocampus* (Sea-horse).

(Günther, *An Introduction to the Study of Fishes*, 1880; Günther, *Catalogue of the Fishes in the Collection of the British Museum*, 1859-70; Cuvier et Valenciennes, *Histoire naturelle des Poissons*, Paris, 1828.)

Fig. 79.—Teleostei. The Cod (*Gadus morrhua*).

ORDER IV.—GANOIDEI.

Sub-ord. 1. Amioidea.

Fam. Amiadæ.—*Amia*.

Sub-ord. 2. Lepidosteoidea.

Fam. *a*. Lepidosteidæ.—*Lepidosteus* (Bony-pike).
Fam. *b*. *Palæoniscidæ.—*Palæoniscus*.
Fam. *c*. *Platysomidæ.—*Platysomus*.
Fam. *d*. *Dapediidæ.—*Dapedius*.
Fam. *e*. *Lepidotidæ.—*Lepidotus*.
Fam. *f*. *Leptolepidæ.—*Leptolepis*.
Fam. *g*. *Pycnodontidæ.—*Pycnodus*.

Sub-ord. 3. Crossopterygidæ.
  Fam. a. Polypteridæ.—*Polypterus, Calamoichthys.*
  Fam. b. *Cœlacanthidæ.—*Cœlacanthus.*
  Fam. c. *Rhombodipteridæ.—*Glyptolœmus, Osteolepis.*
  Fam. d. *Cyclodipteridæ.—*Tristichopterus.*
  Fam. e. *Holoptychiidæ.—*Holoptychius.*
  Fam. f. *Phaneropleuridæ.—*Phaneropleuron.*

Sub-ord. 4. *Acanthophori.
  Fam. Acanthodidæ.—*Acanthodes.*

Sub-ord. 5. *Ostracostei.
  Fam. a. Cephalaspidæ.—*Cephalaspis.*
  Fam. b. Pterichthyidæ.—*Pterichthys.*

Sub-ord. 5. Chondrosteidæ.
  Fam. a. Acipenseridæ. — *Acipenser* (Sturgeon), *Scaphirhynchus.*
  Fam. b. Polyodontidæ.—*Polyodon* (= *Spatularia*, Paddle-fish).

(Joh. Müller, *Ueber Ganoiden und das natürliche System der Fische*, Abhandl. der Berl. Akad., 1848; Agassiz, *Recherches sur les Poissons fossiles*, 1833-43; Huxley, *Essay upon the Systematic Arrangement of the Fishes of the Devonian Epoch*, Mem. Geol. Survey, 1861; and *Illustrations of the Structure of the Crossopterygian Ganoids*, ibid., 1866; Traquair, *The Ganoids of the British Carboniferous Formation*, Palæontograph. Soc., 1877.)

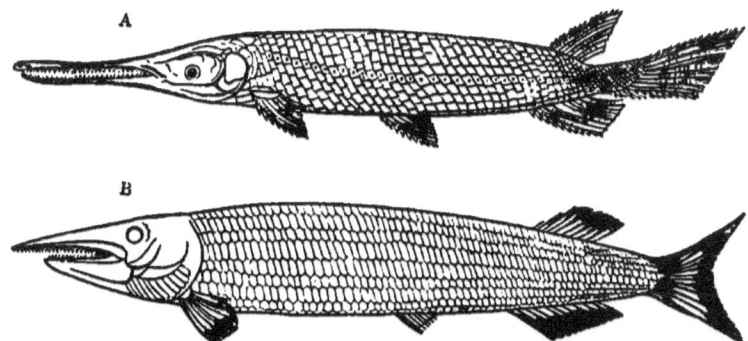

Fig. 80.—A, *Lepidosteus osseus*, the "Gar-Pike" of the American Lakes; B, *Aspidorhynchus*, restored (after Agassiz), a Jurassic Ganoid allied to *Lepidosteus*, but having a homocercal tail.

ORDER V.—ELASMOBRANCHII (CHONDROPTERYGII).

Sub-ord. 1. Holocephali.

Fam. Chimæridæ.—*Chimæra, Callorhynchus.*

Sub-ord. 2. Plagiostomi.

Section A. Selachoidei.

Fam. *a.* Carchariidæ.—*Carcharias* (Shark), *Galeus* (Tope).
Fam. *b.* Lamnidæ.—*Lamna* (Porbeagle), *Carcharodon, Selache* (Basking Shark).
Fam. *c.* Rhinodontidæ.—*Rhinodon.*
Fam. *d.* Notidanidæ.—*Notidanus.*
Fam. *e.* Scylliidæ.—*Scyllium* (Dog-fish).
Fam. *f.* Spinacidæ.—*Spinax, Acanthias* (Piked Dog-fish).
Fam. *g.* Rhinidæ.—*Rhina* (Monk-fish).
Fam. *h.* Pristiophoridæ.—*Pristiophorus.*

Section B. Cestraphori.

Fam. *a.* Cestraciontidæ.—*Cestracion* (Port-Jackson Shark), *Acrodus.
Fam. *b.* *Hybodontidæ.—*Hybodus.*

Section C. Batoidei.

Fam. *a.* Pristidæ.—*Pristis* (Saw-fish).
Fam. *b.* Rhinobatidæ.—*Rhinobatis.*
Fam. *c.* Torpedinidæ.—*Torpedo* (Electric Ray).
Fam. *d.* Raiidæ.—*Raia* (Skate).
Fam. *e.* Trygonidæ.—*Trygon* (Sting-Ray).
Fam. *f.* Myliobatidæ.—*Myliobatis* (Eagle-Ray).

(J. Müller and Henle, *Systematische Beschreibung der Plagiostomen,* Berlin, 1841 ; F. M. Balfour, *Monograph on the Development of the Elasmobranch Fishes,* 1878 ; Günther, *An Introduction to the Study of Fishes,* 1880.)

VERTEBRATA (CHORDATA). 91

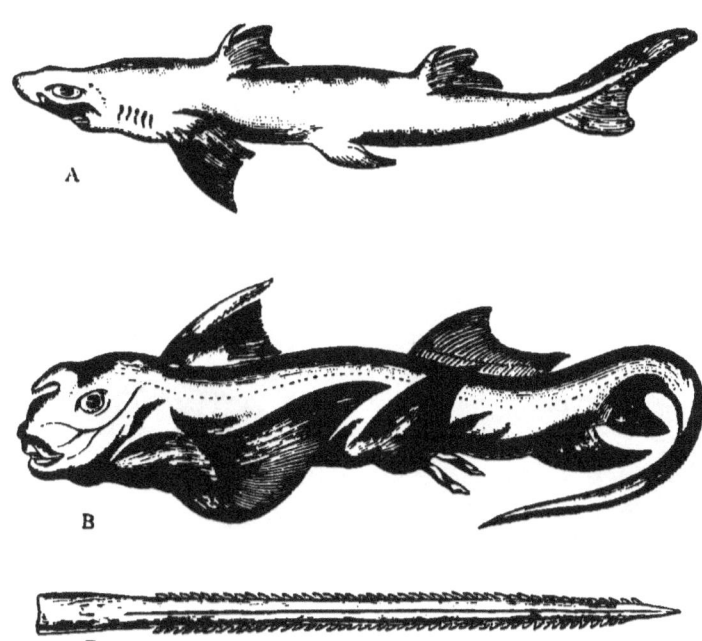

Fig. 81.—A, *Spinax acanthias*, one of the Dog-fishes; B, *Chimæra monstrosa*; C, Tail-spine of an Eagle-Ray (*Myliobatis*).

ORDER VI.—DIPNOI.

Sub-ord. 1. Sirenoidei.

Fam. *a*. Lepidosirenidæ.—*Lepidosiren, Protopterus.*
Fam. *b*. Ceratodidæ.—*Ceratodus* ("Jeevine").

Sub-ord. 2. *Ctenodipterini.—*Dipterus, Ctenodus.*

(Hyrtl, *Lepidosiren paradoxa, Monographie*, Prag, 1845; Owen, *Description of the Lepidosiren annectens*, Trans. Linn. Soc., 1840; Günther, *Description of Ceratodus*, Phil. Trans., 1872; Huxley, *Structure of Ceratodus*, Proc. Zool. Soc., 1876; Pander, *Die Ctenodipterinen des Devonischen Systems*, 1858; Traquair, *On the Genus Dipterus*, &c., Ann. and Mag. Nat. Hist., 1878.)

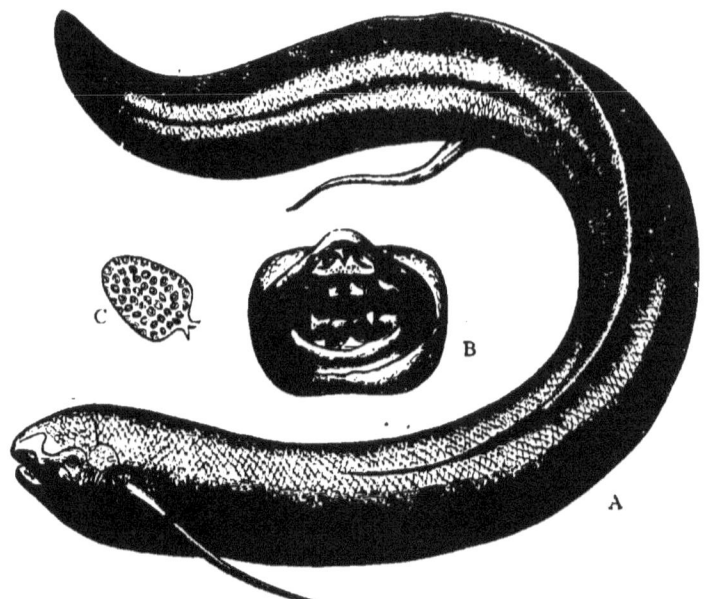

Fig. 82.—A, *Lepidosiren paradoxa*, one of the Mud-fishes; B, Front of the mouth of the same, showing the teeth; C, One of the overlapping scales, enlarged.

CLASS II.—AMPHIBIA (AMPHIBIANS).

ORDER I.—OPHIOMORPHA (PEROMELA).

Fam. Cæciliidæ.—*Cæcilia, Siphonops, Epicrium.*

ORDER II.—URODELA.

Sub-ord. 1. Ichthyodea.

Fam. *a.* Perennibranchiata or Phanerobranchia.—*Siren, Proteus, Menobranchus, Siredon.*
Fam. *b.* Cryptobranchia.—*Amphiuma, Menopoma, Cryptobranchus.*

Sub-ord. 2. Salamandrina.

Fam. *a.* Amblystomidæ.—*Amblystoma, Plethodon.*
Fam. *b.* Salamandriidæ.—*Triton* (Newt), *Salamandra* (Land-Salamander).

The genus _Siredon_ (Axolotl), if regarded as a permanent type, must be placed, as above, among the Perennibranchiate Urodela. It has been shown, however, by Dumeril, Marsh, and others, that under certain circumstances the Axolotl may lose its gills, and may undergo other changes, by which it becomes an _Amblystoma_. There would, therefore, be no impropriety in regarding the ordinary Axolotls as persistent larvæ, and in placing the genus _Siredon_ among the _Amblystomidæ_.

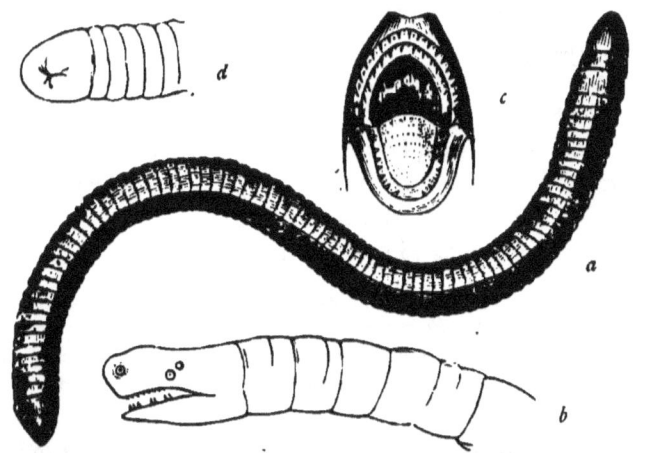

Fig. 83.—Ophiomorpha. a, _Siphonops annulatus_, one of the Cæcilians, much reduced; b, Head; c, Mouth, showing the tongue, teeth, and internal openings of the nostrils; d, Tail and cloacal aperture. (After Dumeril and Bibron.)

Fig. 84.—Tailed Amphibians. A, _Siren lacertina_; B, _Amphiuma_, showing the four minute limbs; c, _Menobranchus maculatus_. (After Mivart.)

ORDER III.—ANOURA.

Sub-ord. 1. Aglossa.
Fam. *a.* Pipidæ.—*Pipa* (Surinam Toad).
Fam. *b.* Dactylethridæ.—*Dactylethra.* (S^a africa)

Sub-ord. 2. Phaneroglossa.
Fam. *a.* Ranidæ (Frogs).—*Rana, Pseudis, Discoglossus, Ceratophrys.*
Fam. *b.* Pelobatidæ.—*Alytes* (Obstetric Toad), *Pelobates, Bombinator.*
Fam. *c.* Bufonidæ (Toads).—*Bufo, Rhinophrynus.*
Fam. *d.* Hylidæ (Tree-frogs).—*Hyla, Notodelphys.*
Fam. *e.* Phyllomedusidæ.—*Phyllomedusa.*
Fam. *f.* Dendrobatidæ.—*Dendrobates.*

The families *Ranidæ, Pelobatidæ,* and *Bufonidæ* are often grouped together as a section of the *Phaneroglossa* under the name of *Oxydactyla*; while the last three families form a section to which the name of *Discodactyla* is given on account of the fact that the toes end in suctorial discs.

ORDER IV.—*LABYRINTHODONTIA.—*Labyrinthodon, Mastodonsaurus, Anthracosaurus, Loxomma, Archegosaurus.*

(Huxley, Article "Amphibia," Encyclopædia Britann., 1875; Günther, *Catalogue of the Batrachia salientia in the Collections of the British Museum,* 1858; Mivart, *The Common Frog,* 1874; Dumeril et Bibron, *Erpétologie générale,* 1834-54; Leydig, *Ueber die Schleichenlurche (Cæciliæ),* Zeitschr. für Wiss. Zool., 1867.)

CLASS III.—REPTILIA (REPTILES).

ORDER I.—CHELONIA.

Fam. *a.* Cheloniidæ (Sea-Turtles).—*Chelone, Sphargis.*
Fam. *b.* Trionycidæ (Soft Tortoises).—*Trionyx.*
Fam. *c.* Chelydidæ.—*Chelys.*
Fam. *d.* Emydidæ. — *Emys* (Terrapin), *Cistudo* (Box-Tortoise).
Fam. *e.* Testudinidæ (Land-Tortoises).—*Testudo, Pyxis.*

(Gray, *Catalogue of the Shield Reptiles in the Collections of the British*

Museum, 1855; Bojanus, *Anatomia testudinis europææ*, 1819-21; Bell, *Monograph of the Testudinata*, Ray Soc., 1836.)

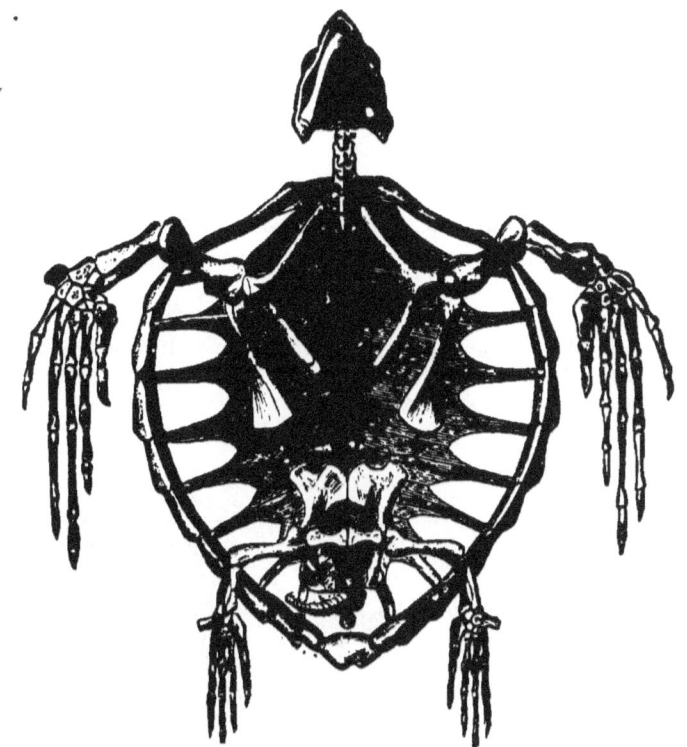

Fig. 85.—Skeleton and carapace of the Logger-headed Turtle (*Chelone caouanna*), viewed from below, the plastron being removed.

ORDER II.—OPHIDIA (SERPENTS).

Section A. Stenostomata.

Fam. *a*. Typhlopidæ.—*Typhlops*.
Fam. *b*. Tortricidæ.—*Tortrix*.
Fam. *c*. Uropeltidæ.—*Uropeltis*.

Section B. Eurystomata.

Sub-ord. 1. Aglyphodontia.

Families: Colubridæ.—*Coluber, Tropidonotus*. Dendrophidæ. — *Dendrophis*. Dipsadidæ. — *Dipsas*. Boidæ.—*Boa, Python, Eunectes*.

Sub-ord. 2. Proteroglypha.

Fam. a. Elapidæ.—*Elaps, Bungarus, Naja* (Cobra), *Ophiophagus* (Hamadryad).

Fam. b. Hydrophidæ (Sea-snakes).—*Hydrophis, Platurus.*

Sub-ord. 3. Solenoglypha.

Fam. a. Viperidæ.—*Vipera, Pelias, Cerastes, Clotho.*

Fam. b. Crotalidæ.—*Crotalus* (Rattle-snake), *Ancistrodon* (Copperhead), *Bothrops* (Fer-de-lance), *Halys, Trimeresurus.*

(Dumeril et Bibron, *Erpétologie générale*, Paris, 1834-45; Günther, *Catalogue of Colubrine Snakes in the Collection of the British Museum*, 1858; Fayrer, *Thanatophidia of India*, 1873.)

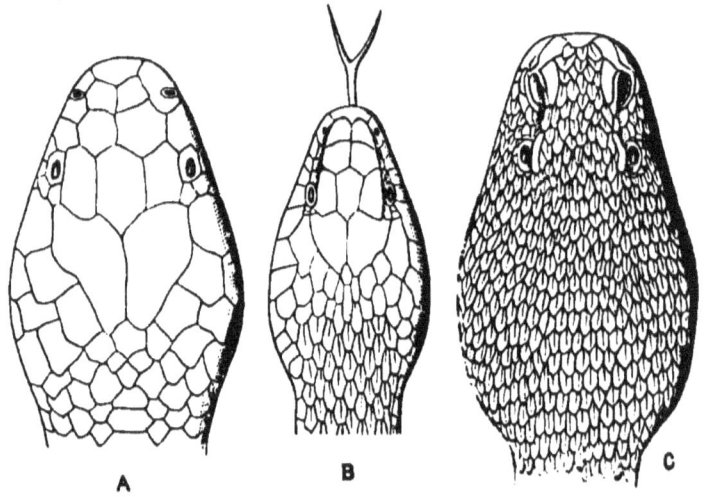

Fig. 86.—Ophidia. A, Head of an Elapine Snake (*Bungarus fasciatus*), viewed from above; B, Head of a Colubrine Snake (*Tropidonotus natrix*); c, Head of a Viperine Snake (*Daboia Russellii*). (A and c are after Sir Joseph Fayrer; B is after Bell.)

ORDER III.—LACERTILIA (LIZARDS).

Sub-ord. 1. Amphisbænoidea (Annulata).

Fam. a. Amphisbænidæ.—*Amphisbæna.*
Fam. b. Chirotidæ.—*Chirotes.*

VERTEBRATA (CHORDATA). 97

Sub-ord. 2. Fissilinguia.
    Fam. *a.* Lacertidæ.—*Lacerta, Zootoca.*
    Fam. *b.* Ameividæ.—*Ameiva, Tejus.*
    Fam. *c.* Varanidæ.—*Varanus* (Monitor).

Sub-ord. 3. Brevilinguia.
    Fam. *a.* Scincoideæ.—*Scincus* (Skink), *Anguis* (Blindworm), *Cyclodus, Seps.*
    Fam. *b.* Chalcididæ.—*Chalcides.*
    Fam. *c.* Zonuridæ.—*Zonurus, Pseudopus* (Sheltopusik).
    Fam. *d.* Geckotidæ (Ascalabotæ).—*Gecko.*
    Fam. *e.* Iguanidæ.—*Iguana, Basiliscus, Draco* (fig. 87).
    Fam. *f.* Agamidæ.—*Agama, Stellio.*

Sub-ord. 4. Vermilinguia.
    Fam. Chamæleontidæ.—*Chamæleo.*

Sub-ord. 5. Rhynchocephalia.
    Fam. Hatteriidæ.—*Hatteria* ("Tuatara").

Sub-ord. 6. *Mosasauria.—*Mosasaurus, Leiodon.*

Sub-ord. 7. *Protorosauria.—*Protorosaurus.*

(Dumeril et Bibron, *Erpétologie générale*, 1834-45 ; Günther, *Anatomy of Hatteria*, Phil. Trans., 1867.)

Fig. 87.—Lacertilia. The "Flying Dragon" (*Draco volans*), viewed from above, of the natural size.

G

ORDER IV.—CROCODILIA.

Sub-ord. 1. Procœlia.
Fam. a. Crocodilidæ.—*Crocodilus* (Crocodile).
Fam. b. Gavialidæ.—*Gavialis* (Gavial).
Fam. c. Alligatoridæ.—*Alligator*.

Sub-ord. 2. *Amphicœlia. — *Belodon, Stagonolepis, Teleosaurus*.

Sub-ord. 3. *Opisthocœlia.—*Streptospondylus*.

(Rathke, *Untersuchungen über die Entwickelung und den Körperbau der Krokodile*, Braunschweig, 1866; Strauch, *Synopsis der gegenwärtig lebenden Crocodile*, Mém. de l'Acad. de St Petersbourg, 1866; Miall, *The Skull of the Crocodile*, 1878; Günther, *The Reptiles of British India*, 1864.)

Fig. 88.—A, Head and anterior portion of the body of *Crocodilus pondicerianus*; B, Hind-foot of the same. (After Günther.)

ORDER V.—*ICHTHYOPTERYGIA.—*Ichthyosaurus*.
ORDER VI.—*SAUROPTERYGIA.—*Plesiosaurus, Pliosaurus*.
ORDER VII.—*ANOMODONTIA.—*Dicynodon, Oudenodon*.
ORDER VIII.—*PTEROSAURIA.—*Pterodactylus, Rhamphorhynchus, Pteranodon*.
ORDER IX.—*DEINOSAURIA.—*Iguanodon, Megalosaurus*.
ORDER X.—*THERIODONTIA.—*Cynodraco*.

VERTEBRATA (CHORDATA). 99

CLASS IV.—AVES (BIRDS).

Sub-class I.—Ratitæ.

Order.—Cursores.

Fam. *a.* Struthionidæ.—*Struthio* (Ostrich).
Fam. *b.* Rheidæ.—*Rhea* (American Ostrich).
Fam. *c.* Dromæidæ.—*Dromaius* (Emeu).
Fam. *d.* Casuariidæ.—*Casuarius* (Cassowary).
Fam. *e.* *Dinornithidæ.—*Dinornis*.
Fam. *f.* *Æpyornithidæ.—*Æpyornis*.
Fam. *g.* Apterygidæ.—*Apteryx* (fig. 89).

(Owen, *Memoir on Dinornis*, Lond., 1866-73; Owen, *Anatomy of the Southern Apteryx*, Trans. Zool. Soc., 1838 and 1842; Owen, *On Dinornis*, Trans. Zool. Soc., 1839-64; Parker, *On the Skull of the Ostrich*, Phil. Trans., 1866.)

Fig. 89.—Cursores. *Apteryx australis*, New Zealand.

Sub-class II:—Carinatæ.

Order I.—Natatores.

Sub-ord. 1. Brevipennatæ.
Fam. *a.* Spheniscidæ (Penguins).—*Spheniscus, Aptenodytes.*

Fam. *b.* Alcidæ.—*Alca* (Auk and Razorbill), *Uria* (Guillemot), *Fratercula* (Puffin).

Fam. *c.* Colymbidæ.—*Colymbus* (Diver), *Podiceps* (Grebe).

Sub-ord. 2. Longipennatæ.

Fam. *a.* Laridæ.—*Larus* (Gull).

Fam. *b.* Sternidæ.—*Sterna* (Tern).

Fam. *c.* Procellaridæ.—*Procellaria* (Fulmar), *Thalassidroma* (Petrel), *Diomedea* (Albatross).

Sub-ord. 3. Totipalmatæ.

Fam. *a.* Pelecanidæ.—*Sula* (Gannet), *Phalacrocorax* (Cormorant), *Pelecanus* (Pelican), *Plotus* (Darter).

Fam. *b.* Tachypetidæ.—*Tachypetes* (Frigate-bird).

Fam. *c.* Phaëtontidæ.—*Phaëton* (Tropic-bird).

Sub-ord 4. Lamellirostres.

Fam. *a.* Anatidæ (Ducks).—*Anas, Fuligula.*

Fam. *b.* Anseridæ (Geese).—*Anser, Cereopsis.*

Fam. *c.* Cygnidæ (Swans).—*Cygnus.*

Fam. *d.* Phœnicopteridæ.—*Phœnicopterus* (Flamingo).

Fig. 90.—Natatores. Penguin (*Aptenodytes patagonica*).

VERTEBRATA (CHORDATA). 101

ORDER II.—GRALLATORES.

Sub-ord. 1. Macrodactyli.

Fam. a. Rallidæ.—*Rallus* (Rail), *Gallinula* (Waterhen).
Fam. b. Parridæ.—*Parra* (Jacana).
Fam. c. Palamedeidæ.—*Palamedea* (Screamer).

Sub-ord. 2. Cultirostres.

Fam. a. Gruidæ.—*Grus* (Crane).
Fam. b. Ardeidæ.—*Ardea* (Heron), *Botaurus* (Bittern).
Fam. c. Tantalidæ.—*Tantalus, Ibis*.
Fam. d. Ciconiidæ.—*Ciconia* (Stork).
Fam. e. Plataleadæ.—*Platalea* (Spoonbill).

Sub-ord. 3. Longirostres.

Fam. Scolopacidæ.—*Scolopax* (Snipe), *Numenius* (Curlew).

Sub-ord. 4. Pressirostres.

Fam. a. Charadriidæ.—*Charadrius* (Plover), *Vanellus* (Lapwing).
Fam. b. Otidæ.—*Otis* (Bustard).

ORDER III.—RASORES.

Sub-ord. 1. Gallinacei (Clamatores).

Fam. a. Tetraonidæ.—*Tetrao* (Grouse), *Lagopus* (Ptarmigan).
Fam. b. Perdicidæ.—*Perdix* (Partridge), *Coturnix* (Quail).
Fam. c. Phasianidæ.—*Phasianus* (Pheasant), *Pavo* (Peafowl), *Meleagris* (Turkey), *Gallus* (Fowl).
Fam. d. Pteroclidæ.—*Pterocles* (Sand-grouse).
Fam. e. Turnicidæ.—*Turnix* (Bush-Quail).
Fam. f. Megapodidæ.—*Megapodius* (Mound-bird).
Fam. g. Cracidæ.—*Crax* (Curassow).
Fam. h. Tinamidæ.—*Tinamus* (Tinamou).
Fam. i. Opisthocomidæ.—*Opisthocomus* (Hoazin).

Sub-ord. 2. Columbacei (Gemitores).

Fam. *a.* Columbidæ (Pigeons).—*Columba, Turtur.*
Fam. *b.* Gouridæ (Ground-pigeons).—*Goura.*
Fam. *c.* Treronidæ (Tree-pigeons).—*Treron.*
Fam. *d.* Didunculidæ.—*Didunculus.*
Fam. *e.* *Dididæ.—*Didus* (Dodo).

ORDER IV.—SCANSORES.

Sub-ord. 1. Cuculiformes.

Fam. *a.* Cuculidæ.—*Cuculus* (Cuckoo), *Coccygus.*
Fam. *b.* Rhamphastidæ.—*Rhamphastos* (Toucan).
Fam. *c.* Musophagidæ.—*Musophaga* (Plantain-eater).
Fam. *d.* Bucconidæ.—*Bucco* (Barbet).
Fam. *e.* Coliidæ.—*Colius.*
Fam. *f.* Trogonidæ.—*Trogon.*

Sub-ord. 2. Piciformes.

Fam. *a.* Picidæ (Woodpeckers).—*Picus, Colaptes.*
Fam. *b.* Yungidæ.—*Yunx* (Wryneck).

Sub-ord. 3. Psittaciformes.

Fam. *a.* Psittacidæ (Parrots).—*Psittacus, Agapornis.*
Fam. *b.* Plyctolophidæ (Cockatoos).—*Plyctolophus.*
Fam. *c.* Macrocercidæ (Macaws).—*Macrocercus (Ara), Pezoporus.*
Fam. *d.* Trichoglossidæ.—*Trichoglossus* (Parrakeet), *Lorius* (Lory), *Nestor.*
Fam. *e.* Strigopidæ.—*Strigops* (Kakapo).

ORDER V.—INSESSORES (PASSERES).

Sub-ord. 1. Conirostres.

Families: Corvidæ.—*Corvus* (Crow), *Pica* (Magpie), *Garrulus* (Jay). Sturnidæ.—*Sturnus* (Starling), *Gracula.* Paradiseidæ (Birds of Paradise).—*Paradisea.* Oriolidæ.—*Oriolus.* Ampelidæ (Chatterers).—*Ampelis.* Fringillidæ (Finches). — *Passer* (Sparrow), *Emberiza* (Bunting). Tanagridæ.—*Tanagra.*

VERTEBRATA (CHORDATA). 103

Fig. 91.—Scansores. The Owl-Parrot (*Strigops habroptilus*), New Zealand.

Sub-ord. 2. Dentirostres.
Families : Muscicapidæ. — *Muscicapa* (Fly-catcher). Laniidæ. — *Lanius* (Shrike). Sylviadæ. — *Sylvia* (Warbler), *Saxicola* (Stone-chat). Turdidæ.—*Turdus* (Thrush). Motacillidæ.—*Motacilla* (Wagtail), *Anthus* (Titlark). Troglodytidæ.—*Troglodytes* (Wren). Paridæ.—*Parus* (Titmouse). Tyrannidæ.—*Tyrannus*.

Sub-ord. 3.—Tenuirostres.
Families : Promeropidæ.—*Promerops*, *Nectarinia* (Sunbird). Meliphagidæ. — *Meliphaga* (Honey-eater). Certhiidæ. — *Certhia* (Creeper). Sittidæ. — *Sitta* (Nuthatch). Upupidæ.—*Upupa* (Hoopoe). Trochilidæ (Humming-birds).—*Trochilus*.

Sub-ord. 4. Fissirostres.
Families : Hirundinidæ (Swallows).—*Hirundo*. Cypselidæ (Swifts). — *Cypselus*, *Collocalia*. Caprimulgidæ (Goat-suckers).—*Caprimulgus*, *Steatornis*. Meropidæ (Bee-eaters).—*Merops*. Alcedinidæ (Kingfishers).—*Alcedo*, *Dacelo*.

104 CLASSIFICATION OF THE ANIMAL KINGDOM.

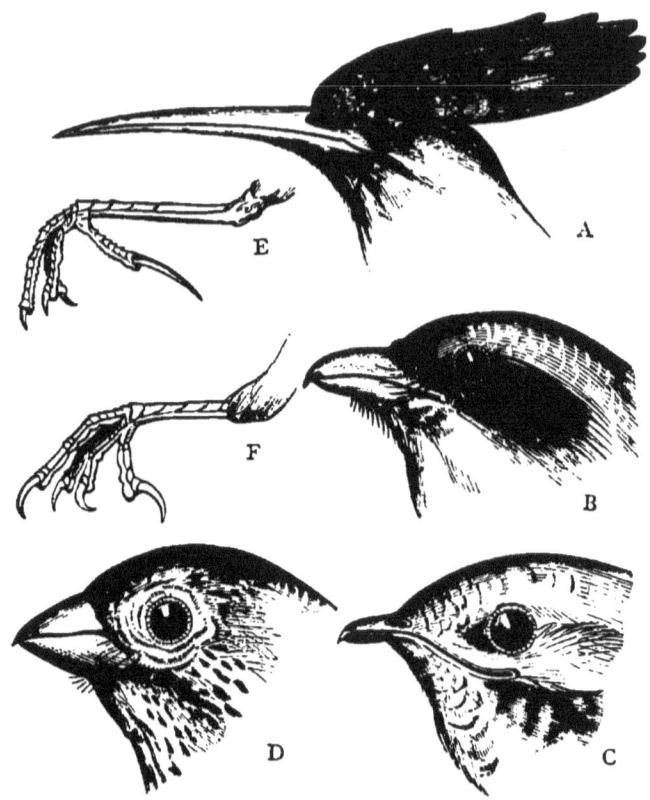

Fig. 92.—Insessores. A, Head of Hoopoe (*Upupa epops*), showing the Tenuirostral type of beak; B, Head of Red-backed Shrike (*Lanius collurio*), showing the Dentirostral type of beak; C, Head of White-bellied Swift (*Cypselus melba*), showing the Fissirostral type of beak; D, Head of Corn-Bunting (*Emberiza miliaria*), showing the Conirostral type of beak; E, Foot of the Yellow Wagtail (*Motacilla sulphurea*); F, Foot of a Finch (*Fringilla*).

ORDER VI.—RAPTORES (BIRDS OF PREY).

Fam. *a*. Strigidæ (Owls).—*Strix, Scops, Bubo, Athene*.
Fam. *b*. Falconidæ.—*Aquila* (Eagle), *Falco* (Falcon), *Buteo* (Buzzard), *Milvus* (Kite).
Fam. *c*. Vulturidæ (Vultures).—*Neophron, Gypaëtus*.
Fam. *d*. Cathartidæ (American Vultures).—*Cathartes* (Californian Vulture), *Sarcorhamphus* (Condor).
Fam. *e*. Gypogeranidæ.—*Gypogeranus* (Secretary Vulture).

(Gray, *Genera of Birds*, London, 1849; Newton, Article "*Birds*," Encyclopædia Brit., 9th ed., 1875; Owen, Article "*Aves*," Todd's Cyclopædia Anat. and Phys., 1836; Selenka, *Aves*, Bronn's Klassen und Ordnungen des Thierreichs, 1869-82; Huxley, *On the Classification of Birds*, Proc. Zool. Soc., 1867; Tiedemann, *Anatomie und Naturgeschichte der Vögel*, Heidelberg, 1848; Eyton, *Osteologia Avium*, London, 1858-60; Macgillivray, *History of British Birds*, 1839-41.)

SUB-CLASS III.—*SAURORNITHES.

ORDER.—SAURURÆ.—*Archæopteryx.*

SUB-CLASS IV.—*ODONTORNITHES.

ORDER I.—ODONTOLCÆ.—*Hesperornis.*
ORDER II.—ODONTOTORMÆ.—*Ichthyornis, Apatornis.*

(Owen, *Archæopteryx macrura*, Phil. Trans., 1863; Carl Vogt, *Archæopteryx*, Revue Scientifique, 1879; Marsh, *Jurassic Birds and their Allies*, Rep. Brit. Ass., 1881; Marsh, *Odontornithes, a Monograph of the Extinct Toothed Birds of North America*, New Haven, 1880.)

Fig. 93.—Odontornithes. Skeleton of *Hesperornis regalis*, restored. (After Marsh.) About one-tenth of the natural size.

106 CLASSIFICATION OF THE ANIMAL KINGDOM.

CLASS V.—MAMMALIA (QUADRUPEDS).

SUB-CLASS I.—ORNITHODELPHIA.

ORDER I.—MONOTREMATA.

Fam. *a.* Ornithorhynchidæ.—*Ornithorhynchus* (Duck-mole, fig. 94).

Fam. *b.* Echidnidæ.—*Echidna.* 4 species, 1 of wh. is f

(Owen, Art. "*Monotremata*," Todd's Cyclopædia Anat. and Phys., 1841 ; Meckel, *Ornithorhynchi paradoxi descriptio anatomica*, 1826 ; Waterhouse, *A Natural History of the Mammalia*, vol. i., London, 1846.)

Fig. 94.—Monotremata. *Ornithorhynchus anatinus*, Australia.

SUB-CLASS II.—DIDELPHIA.

ORDER I.—MARSUPIALIA.

Sub-ord. 1. Diprotodontia.

Section Rhizophaga.

Fam. Phascolomydæ.—*Phascolomys* (Wombat).

VERTEBRATA (CHORDATA). 107

Section Poëphaga.
  Fam. Macropodidæ.—*Macropus* (Kangaroo), *Dendrolagus* (Tree-kangaroo), *Hypsiprymnus* (Kangaroo-rat).

Section Carpophaga.
  Fam. *a.* Phascolarctidæ.—*Phascolarctos* (Koala).
  Fam. *b.* Phalangistidæ (Phalangers).—*Phalangista, Petaurus.*

Sub-ord. 2. Polyprotodontia.

Section Entomophaga.
  Fam. *a.* Peramelidæ.—*Perameles* (Bandicoot), *Chœropus.*
  Fam. *b.* Didelphidæ.—*Didelphys* (Opossum), *Cheironectes.*

Section Sarcophaga.
  Fam. Dasyuridæ.—*Thylacinus, Dasyurus.*

(Waterhouse, *A Natural History of the Mammalia*, vol. i., London, 1846; Owen, *Classification of the Marsupialia*, Trans. Zool. Soc., 1839; Owen, *Fossil Mammalia of Australia*, 1877; Gould, *The Mammals of Australia*, 1863.)

Fig. 95.—Marsupialia. The female of *Didelphys dorsigera*, one of the South American Opossums, carrying its young upon its back.

108 CLASSIFICATION OF THE ANIMAL KINGDOM.

SUB-CLASS III.—MONODELPHIA.

ORDER I.—EDENTATA (BRUTA).
  Sub-ord. 1. Tardigrada.
    Fam. Bradypodidæ (Sloths).—*Bradypus, Cholœpus.*
  Sub-ord. 2. *Gravigrada.
    Fam. *Megatheridæ. — *Megatherium, Mylodon, Megalonyx.*
  Sub-ord. 3. Loricata.
    Fam. *a.* Dasypodidæ (Armadillos).—*Dasypus, Chlamyphorus.*
    Fam. *b.* *Glyptodontidæ.—*Glyptodon.*
  Sub-ord. 4. Vermilinguia.
    Fam. *a.* Myrmecophagidæ.—*Myrmecophaga* (Ant-eater), *Cyclothurus.*
    Fam. *b.* Manidæ.—*Manis* (Pangolin).
    Fam. *c.* Orycteropidæ.—*Orycteropus* (Aardvark).

Fig. 96.—Edentata. The three-banded Armadillo (*Tolypeutes conurus*), one-third of the natural size. (After Murie.)

(Rapp, *Anatomische Untersuchungen über die Edentaten*, Tübingen, 1852; Turner, *Classification of the Edentata*, Proc. Zool. Soc., 1851; Jäger, *An-*

atomische Untersuchung des Orycteropus Capensis, Stuttgart, 1837 ; Owen, *Memoir on the Megatherium*, 1860 ; Murie, *On the Habits, Structure, and Relations of the Three-banded Armadillo*, Trans. Linn. Soc., 1872 ; Owen, *On the Anatomy of the Great Ant-Eater*, Trans. Zool. Soc., 1856-57.)

ORDER II.—SIRENIA.

Fam. Manatidæ (Sea-cows).—*Manatus* (Manatee), *Halicore* (Dugong), *Rhytina* (extinct within the historical period), **Halitherium*.

(Murie, *Form and Structure of the Manatee*, Trans. Zool. Soc., 1872 ; Owen, *Anatomy of the Dugong*, Proc. Zool. Soc., 1838.)

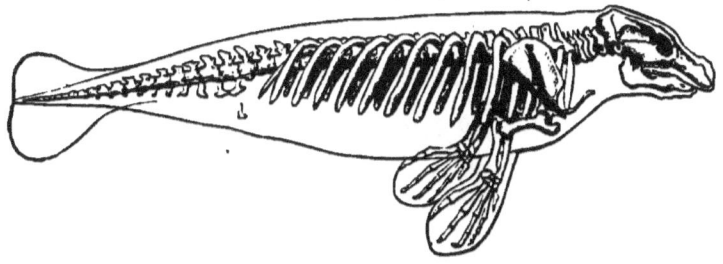

Fig. 97.—Sirenia. Skeleton of the Manatee (*Manatus Americanus*).

ORDER III.—CETACEA.

Sub-ord. 1. Mysticeti.

Fam. Balænidæ.—*Balæna* (Right Whale), *Balænoptera* (Rorqual).

Sub-ord. 2. Odontoceti.

Fam. *a*. Catodontidæ.—*Physeter* (Sperm-whale).
Fam. *b*. Delphinidæ. — *Delphinus* (Dolphin), *Phocæna* (Porpoise), *Platanista*.
Fam. *c*. Monodontidæ.—*Monodon* (Narwhal).
Fam. *d*. Rhynchoceti.—*Ziphius, Hyperoödon*.
Fam. *e*. *Zeuglodontidæ.—*Zeuglodon, Squalodon*.

(F. Cuvier, *Histoire naturelle des Cétacés*, Paris, 1836 ; Eschricht, *Untersuchungen über die nordischen Walthiere*, Leipzig, 1849 ; Gray, *Synopsis of the Species of Whales and Dolphins in the British Museum*, 1868 ; Flower, *Notes on the Skeletons of Whales*, Proc. Zool. Soc., 1864.)

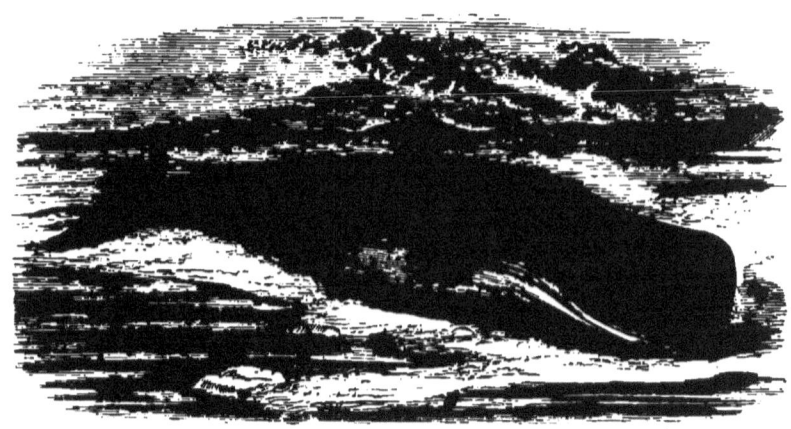

Fig. 98.—Cetacea. Spermaceti Whale (*Physeter macrocephalus*).

ORDER IV.—UNGULATA.

Section Perissodactyla (Odd-toed Ungulates).

Fam. *a.* *Coryphodontia.—*Coryphodon.*
Fam. *b.* Rhinocerotidæ.—*Rhinoceros.*
Fam. *c.* Tapiridæ.—*Tapirus* (Tapir).
Fam. *d.* *Brontotheridæ.—*Brontotherium.*
Fam. *e.* *Palæotheridæ.—*Palæotherium.*
Fam. *f.* *Macrauchenidæ.—*Macrauchenia.*
Fam. *g.* Equidæ.—*Equus* (Horse), *Asinus* (Ass, Zebra), **Orohippus*, **Hipparion.*

Section Artiodactyla (Even-toed Ungulates).

Sub-ord. 1. Omnivora (Bunodonta).

Fam. *a.* Hippopotamidæ.—*Hippopotamus.*
Fam. *b.* Suidæ.—*Sus* (Pig), *Dicotyles* (Peccary), *Phacochœrus* (Wart-hog).
Fam. *c.* *Anoplotheridæ.—*Anoplotherium.*
Fam. *d.* *Oreodontidæ.—*Oreodon.*

Fig. 99.—Grinding surface of the molar and premolar teeth of a Peccary (*Dicotyles labiatus*), showing the bunodont type of dentition. (After Giebel.)

VERTEBRATA (CHORDATA).

Fig. 100.—Grinding surface of the molar and præmolar teeth of the Giraffe (*Camelopardalis Giraffa*), showing the selenodont type of dentition.

Sub-ord. 2. Ruminantia (Selenodonta).

Fam. *a*. Camelidæ (Tylopoda). — *Camelus* (Camel), *Auchenia* (Llama).
Fam. *b*. Tragulidæ (Chevrotains).—*Tragulus*, *Hyomoschus*.
Fam. *c*. Cervidæ (Deer).—*Cervus*, *Dama*, *Alces*.
Fam. *d*. Camelopardalidæ.—*Camelopardalis* (Giraffe).
Fam. *e*. Antilopidæ (Antelopes). — *Rupicapra*, *Antilocapra*.
Fam. *f*. Ovidæ.—*Ovis* (Sheep), *Capra* (Goat).
Fam. *g*. Bovidæ.—*Bos* (Ox), *Ovibos* (Musk-ox), *Bubalus* (Buffalo).

(Cuvier, *Recherches sur les Ossemens fossiles*, Paris, 1846; Giebel, *Die Saügethiere in zoologischer, anatomischer, und palæontologischer Beziehung*, Leipzig, 1859; Flower, *Osteology of the Mammalia*, 1876; Owen, *Anatomy and Physiology of Vertebrated Animals*, vols. ii. and iii., 1866 and 1868; Pander and D'Alton, *Die Skelete der Wiederkaüer;* Rütimeyer, *Versuch einer natürlichen Geschichte des Rindes in seinen Beziehungen zu den Wiederkaüern im Allgemeinen*, 1866; Nathusius, *Die Racen des Schweines*, Berlin, 1860.)

ORDER V.—*DINOCERATA.—*Dinoceras*.

(Marsh, *Principal Characters of the Dinocerata*, Amer. Journ. Sci. and Arts, 1876.)

ORDER VI.—*TILLODONTIA.—*Tillotherium*.

(Marsh, *Principal Characters of the Tillodontia*, Amer. Journ. Sci. and Arts, 1876.)

112 CLASSIFICATION OF THE ANIMAL KINGDOM.

ORDER VII.—*TOXODONTIA.—*Toxodon.*

(Owen, *Fossil Mammalia of the Voyage of the Beagle*, 1840.)

ORDER VIII.—HYRACOIDEA.

Fam. Hyracidæ.—*Hyrax.*

(Owen, *On the Anatomy of the Cape Hyrax*, Proc. Zool. Soc., 1832; Schreber, *Naturgeschichte der Saügethiere*, 1775-1855.)

ORDER IX.—PROBOSCIDEA.

Fam. *a.* Elephantidæ.—*Elephas,* *Mastodon.*
Fam. *b.* *Deinotheridæ.—*Deinotherium.*

(Cuvier, *Recherches sur les Ossemens fossiles*, Paris, 1846; Mayer, *Beiträge zur anatomie des Elephanten*, Nova Acta, 1847; Kaup, *Deinotherium giganteum*, Isis, Bd. IV., 1829; Falconer, *Palæontological Memoirs*, 1868; Leith Adams, *Monograph of the British Fossil Elephants*, Palæontograph. Soc., 1877-78.)

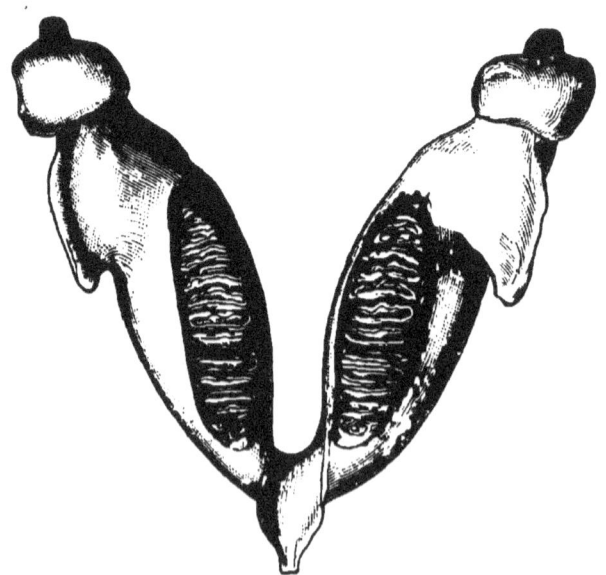

Fig. 101.—Lower jaw of the Indian Elephant (*Elephas Indicus*), viewed from above, showing the molar teeth. Greatly reduced in size.

ORDER X.—CARNIVORA (FERÆ).

Sub-ord. 1. Pinnipedia.

Fam. *a.* Phocidæ (Seals).—*Phoca, Cystophora* (Sea-elephant), *Halichœrus.*
Fam. *b.* Otariidæ (Eared-seals).—*Otaria.*
Fam. *c.* Trichecidæ.—*Trichecus* (Walrus).

Sub-ord. 2. Plantigrada.

Fam. *a.* Ursidæ (Bears).—*Ursus, Helarctos* (Sun-bear).
Fam. *b.* Procyonidæ.—*Procyon* (Racoon), *Nasua* (Coati, fig. 102), *Bassaris.*
Fam. *c.* Cercoleptidæ.—*Cercoleptes* (Kinkajou).
Fam. *d.* Æluridæ.—*Ælurus* (" Panda ").
Fam. *e.* Melidæ (Badgers).—*Meles, Mellivora* (Honey-badger).

Sub-ord. 3. Digitigrada.

Fam. *a.* Mustelidæ.—*Putorius* (Weasel and Polecat), *Mustela* (Marten), *Lutra* (Otter).
Fam. *b.* Viverridæ. — *Viverra* (Civet-cat), *Herpestes* (Ichneumon).
Fam. *c.* Hyænidæ.—*Hyæna, Proteles* (Aardwolf).
Fam. *d.* Canidæ.—*Canis* (Dog, Wolf, &c.), *Vulpes* (Fox).
Fam. *e.* Felidæ.—*Felis* (Cat, Lion, &c.), *Lynx* (Lynx), *Machairodus.*

On the classification adopted by Prof. Flower, the *Carnivora* are divided into the three following sections, the *Pinnipedia* being left out: (1), ARCTOIDEA, comprising the *Ursidæ, Procyonidæ, Æluridæ,* and *Mustelidæ.* (2) CYNOIDEA, comprising only the family of the *Canidæ.* (3) ÆLUROIDEA, comprising the *Viverridæ, Hyænidæ, Cryptoproctidæ,* and *Felidæ.*

(Strauss-Durckheim, *Anatomie descriptive et comparative du Chat*, Paris 1845; St. George Mivart, *The Cat*, 1881; Flower, *On the Value of the Characters of the Base of the Cranium in the Classification of the Order Carnivora*, Proc. Zool. Soc., 1869; Allen, *Monograph of the North American Pinnipeds*, Washington, 1880; Bell, Art. " *Carnivora,*" Todd's Cyclopædia Anat. and Phys., 1835.)

114 CLASSIFICATION OF THE ANIMAL KINGDOM.

Fig. 102.—Carnivora. *Nasua fusca*, the Brown Coati.

Order XI.—Rodentia (Glires).

Sub-ord. 1. Duplicidentata.
   Fam. *a*. Leporidæ.—*Lepus* (Hare and Rabbit).
   Fam. *b*. Lagomydæ.—*Lagomys* (Pika).

Sub-ord. 2. Simplicidentata.
   Fam. *a*. Caviidæ.—*Cavia* (Cavy), *Hydrochœrus* (Capybara), *Dasyprocta* (Agouti).
   Fam. *b*. Hystricidæ.—*Hystrix* (Porcupine).
   Fam. *c*. Cercolabidæ.—*Cercolabes*.
   Fam. *d*. Octodontidæ.—*Octodon, Ctenomys, Myopotamus* (Coypu).
   Fam. *e*. Chinchillidæ.—*Chinchilla*.
   Fam. *f*. Castoridæ.—*Castor* (Beaver).
   Fam. *g*. Saccomydæ.—*Geomys* (Gopher).
   Fam. *h*. Spalacidæ.—*Spalax* (Mole-rat).
   Fam. *i*. Muridæ.—*Mus* (Rat, Mouse), *Myodes* (Lemming).

VERTEBRATA (CHORDATA). 115

Fam. *j.* Dipodidæ.—*Dipus* (Jerboa).
Fam. *k.* Myoxidæ.—*Myoxus* (Dormouse).
Fam. *l.* Sciuridæ.—*Sciurus* (Squirrel), *Arctomys* (Marmot).

Sub-ord. 3. Hebedidentata.
Fam. *Mesotheriidæ.—*Mesotherium*.

(Waterhouse, *A Natural History of the Mammalia*, vol. ii., Rodentia, London, 1838 ; Lilljeborg, *Systematisk Öfversigt af de gnagande Däggjuren, Glires*, 1866 ; Alston, *On the Classification of the Order Glires*, Proc. Zool. Soc., 1876 ; Coues and Allen, *Monographs of the North American Rodentia* Washington, 1877.)

Fig. 103.—Rodentia. The Agouti (*Dasyprocta aguti*).

ORDER XII. CHEIROPTERA (BATS).

Sub-ord. 1. Insectivora (Microcheiroptera).
Fam. *a.* Vespertilionidæ.—*Vespertilio, Plecotus*.
Fam. *b.* Rhinolophidæ.—*Rhinolophus* (Horse-shoe Bat).
Fam. *c.* Noctilionidæ.—*Emballonura*.
Fam. *d.* Phyllostomidæ.—*Phyllostoma*.

116   CLASSIFICATION OF THE ANIMAL KINGDOM.

Sub-ord. 2. Frugivora (Megacheiroptera).
Fam. Pteropidæ.—*Pteropus* (Fox-bat).

(Dobson, *Monograph of the Asiatic Cheiroptera*, 1876; Bell, Art. "*Cheiroptera*," Todd's Cyclopædia Anat. and Phys., 1835; Blasius, *Naturgeschichte der Saügethiere Deutschlands*, Braunschweig, 1857.)

Fig. 104.—Cheiroptera. *Vespertilio discolor*, one-half the natural size.

ORDER XIII.—INSECTIVORA.

Fam. *a*. Talpidæ.—*Talpa* (Mole), *Condylura*, *Chrysochloris* (Golden-mole).
Fam. *b*. Potamogalidæ.—*Potamogale*.
Fam. *c*. Soricidæ.—*Sorex* (Shrew-mouse), *Myogale* (Desman).
Fam. *d*. Erinaceidæ.—*Erinaceus* (Hedgehog).
Fam. *e*. Centetidæ.—*Centetes*, *Ericulus*.
Fam. *f*. Tupaiidæ.—*Tupaia* (Banxring).
Fam. *g*. Macroscelidæ.—*Macrosceles* (Elephant-shrew).
Fam. *h*. Galeopithecidæ.—*Galeopithecus* (Flying-lemur).

(Mivart, *Classification of the Insectivora*, Proc. Zool. Soc., 1871—also on the *Osteology of the Insectivora*, Journ. Anat. and Phys., vol. ii.; Allman,

VERTEBRATA (CHORDATA). 117

On the Characters and Affinities of Potamogale, Trans. Zool. Soc., 1866;
Fitzinger, *Ueber die Naturliche Familie der Igel (Erinacei)*, Sitzb. der K.
Akad. Wiss. Wien, 1867, with papers on the *Macroscelidæ* and *Soricidæ;*
Fitzinger, *Die natürliche Familie der Maulwürfe (Talpæ)*, ibid., 1869;
Gill, *Synopsis of Insectivorous Mammalia*, Bull. U.S. Geol. Survey, 1875;
Coues, *Precursory Notes on American Insectivorous Mammals*, Bull. U.S.
Geol. Survey, 1877.)

Fig. 105.—Insectivora. The Hedgehog (*Erinaceus Europæus*).

ORDER XIV.—QUADRUMANA (MONKEYS).

Sub-ord. 1. Strepsirhina (Prosimiæ).

Fam. *a.* Cheiromydæ.—*Cheiromys* (Aye-Aye).
Fam. *b.* Tarsiidæ.—*Tarsius.*
Fam. *c.* Nycticebidæ.—*Nycticebus, Stenops.*
Fam. *d.* Lemuridæ.—*Lemur, Indris* (Indri), *Galago.*

Sub-ord. 2. Platyrhina.

Fam. *a.* Hapalidæ.—*Hapale* (Marmoset), *Midas.*
Fam. *b.* Cebidæ.—*Cebus* (Capuchin-monkey), *Ateles*
(Spider-monkey), *Mycetes* (Howler).

118 CLASSIFICATION OF THE ANIMAL KINGDOM.

Sub-ord. 3. Catarhina.

Fam. *a.* Semnopithecidæ.—*Semnopithecus, Cercopithecus, Macacus* (Macaque), *Rhesus.*

Fam. *b.* Cynocephalidæ (Baboons).—*Cynocephalus.*

Fam. *c.* Simiidæ.—*Hylobates* (Gibbon), *Troglodytes* (Gorilla, Chimpanzee), *Simia* (Orang).

(Mivart, Art. "*Apes,*" Encyclo. Brit., 9th ed., 1875; Mivart, *Appendicular Skeleton of Primates,* Phil. Trans., 1867; Mivart, *The Zoological Rank of the Lemuroidea,* Proc. Zool. Soc., 1873; Owen, *Monograph on the Aye-Aye,* 1863; Owen, *Memoir on the Gorilla,* 1865; Vrolik, *Recherches sur le Chimpansé,* 1841.)

Fig. 106.—Quadrumana. *Cercocebus mona*, one-seventh of the natural size.

ORDER XV.—BIMANA.—*Homo* (Man).

(Pritchard, *Natural History of Man,* 1843; Darwin, *Descent of Man,* 1871; Nott and Gliddon, *Types of Mankind,* 1854; Tylor, *Anthropology, an Introduction to the Study of Man and Civilisation,* 1881; Lubbock, *Prehistoric Times,* 2d ed., 1869.)

INDEX OF GENERA.

(Extinct Genera are marked with an asterisk.)

Acanthella, 10.
Acanthias, 90.
Acanthobdella, 45.
*Acanthodes, 89.
Acanthometra, 7.
Acarus, 61.
Acltheres, 48.
Acicula, 82.
Acineta, 8.
Acipenser, 89.
Acmœa, 81.
Acridium, 67.
*Acrodus, 90.
Actinia, 21.
Actinocrinus, 32.
Actinometra, 32.
Actinosphærium, 8.
Ædipoda, 67.
Æga, 56.
Ægina, 17.
Ælurus, 113.
Æolis, 81.
Æolosoma, 46.
Æpyornis, 99.
Æquorea, 17.
Æshna, 68.
Agalma, 20.
Agama, 97.
Agapornis, 102.
*Agnostus, 54.
Agrion, 68.
Albertia, 42.
Alca, 100.
Alcedo, 103.
Alces, 111.
Alcyonidium, 74.
Alcyonium, 22.
Alitta, 46.
Alligator, 98.
Alpheus, 58.

*Alveolites, 22.
Alveopora, 22.
Alytes, 94.
Amaroucium, 76.
Amblystoma, 92.
Ameiva, 97.
Amia, 88.
Ammodytes, 87.
Ammonites, 84.
Amœba, 4, 5.
Ampelis, 102.
Amphicora, 46.
Amphidetus, 28.
Amphihelia, 22.
Amphileptus, 8, 9.
Amphinome, 46.
Amphioxus, 86.
Amphisbœna, 96.
Amphitrite, 46.
Amphiuma, 92.
Ampullaria, 81.
*Ananchytes, 28.
Anapta, 31.
Anarrhicas, 87.
Anas, 100.
Anatina, 79.
Anceus, 56.
Ancistrodon, 96.
Andrena, 71.
Androctonus, 62.
Anguis, 97.
Anodon, 78.
Anomia, 78.
*Anoplotherium, 110.
Anser, 100.
Antedon, 32, 33.
Antennularia, 17.
*Anthocrinus, 32.
Anthomyia, 69.
*Anthracosaurus, 94.

INDEX OF GENERA.

Anthura, 56.
Anthus, 103.
Antilocapra, 111.
Antipathes, 22.
*Apatornis, 105.
Aphis, 66.
Aphrodite, 46.
Aphrophora, 66.
*Apiocrinus, 32.
Apis, 71.
Aplysia, 81.
Aplysina, 9.
Aporrhais, 81.
Appendicularia, 75.
Apsilus, 42.
Aptenodytes, 99.
Apteryx, 99.
*Aptychopsis, 53.
Apus, 53.
Aquila, 104.
Ara, 102.
Arachnactis, 22.
Arbacia, 27.
Arca, 78.
Arcella, 4.
*Archæocidaris, 27.
*Archæopteryx, 105.
Archaster, 29.
*Archegosaurus, 94.
*Archiulus, 64.
Arctomys, 115.
Arcturus, 56.
Ardea, 101.
Arenicola, 46.
Argiope, 77.
Argonauta, 84.
Armadillo, 57.
Artemia, 53.
Artemis, 79.
*Asaphus, 54.
Ascaris, 41.
Ascidia, 75.
Asellus, 56.
Asilus, 69.
Asinus, 110.
Aspergillum, 79.
Aspidodiadema, 27.
Aspidisca, 8.
Asplanchna, 42.
Astacus, 58.
Astarte, 79.
Asterocanthion, 29.
Asterias, 29.
Asterina, 29.
Asteronyx, 30.
Asterophyton, 30.
Asthenosoma, 27.
Astræa, 22.
Astropecten, 29.
Astrorhiza, 5.
*Athyris, 77.
Atlanta, 82.
Auchenia, 111.
*Aulocopium, 10.

*Aulopora, 25.
Aurelia, 20.
Auricula, 82.
Autolytus, 46.
Avicula, 78.
Axinella, 10.

*Bactrites, 84.
*Baculites, 84.
Balæna, 109.
Balænoptera, 109.
Balanoglossus, 47.
Balanophyllia, 22.
Balanus, 49.
Balatro, 42.
Balistes, 88.
Basiliscus, 97.
Bassaris, 113.
Bathycrinus, 32.
Bathygadus, 87.
Bdella, 61.
*Belemnitella, 84.
*Belemnites, 84.
*Belodon, 98.
Beroë, 25.
Birgus, 59.
Blatta, 67.
Boa, 95.
Bolina, 25.
Bombinator, 94.
Bombus, 71.
Bombyx, 70.
Bonellia, 42.
Bopyrus, 57.
Borlasia, 39.
Bosmina, 52, 53.
Botaurus, 101.
Bothriocephalus, 36.
Bothrops, 96.
Botryllus, 76.
Bougainvillea, 16.
*Bourgueticrinus, 32.
Brachionus, 42.
Bradypus, 108.
Branchiobdella, 45.
Branchipus, 53.
Brisinga, 29.
Brissus, 28.
*Brontotherium, 110.
Bubalus, 111.
Bubo, 104.
Buccinum, 81.
Bucco, 102.
Bufo, 94.
Bugula, 74.
Bulimina, 6.
Bulimus, 82.
Bulla, 81.
Bungarus, 96.
Bursaria, 8, 9.
Buteo, 104.
Buthus, 62.

Cæcilia, 92.

INDEX OF GENERA.

Calamoichthys, 89.
Calanus, 51.
Calappa, 59.
Caligus, 48.
Callianassa, 58.
Callorhynchus, 90.
Calymene, 54.
Calyptræa, 81.
Camelopardalis, 111.
Camelus, 111.
Campanularia, 17.
Campodea, 66.
Cancer, 59.
Candona, 50.
Canis, 113.
Capra, 111.
Caprella, 55.
Caprimulgus, 103.
Capsus, 66.
Capulus, 81.
Carcharias, 90.
Carcharodon, 90.
Carcinus, 59.
Cardita, 79.
Cardium, 79.
Carduella, 20.
Carinaria, 82.
Carpilius, 59.
Caryocrinus, 35.
Caryophyllæus, 36.
Cassidulina, 6.
Cassiopeia, 20.
Castalia, 46.
Castor, 114.
Casuarius, 99.
Catenicella, 74.
Cathartes, 104.
Cavia, 114.
Cebus, 117.
Cecidomyia, 69.
Cellaria, 74.
Cellepora, 74.
Cellularia, 74.
Centetes, 116.
Cephalaspis, 89.
Cephea, 20.
Cerastes, 96.
Ceratites, 84.
Ceratium, 8.
Ceratodus, 91.
Ceratophrys, 94.
Cercolabes, 114.
Cercoleptes, 113.
Cercomonas, 8.
Cercopithecus, 118.
Cereopsis, 100.
Cerithium, 81.
Certhia, 103.
Cervus, 111.
Cestracion, 90.
Cestum, 25.
Cetonia, 72.
Chæropus, 107.
Chætetes, 25.

Chætogaster, 46.
Chætonotus, 42.
Chætosoma, 41.
Chalcides, 97.
Chalcis, 71.
Chalina, 9.
Chama, 79.
Chamæleo, 97.
Charadrius, 101.
Cheiracanthus, 41.
Cheiromys, 117.
Cheironectes, 107.
Chelifer, 62.
Chelone, 94.
Chelura, 55.
Chelys, 94.
Chemnitzia, 81.
Chermes, 66.
Chernes, 62.
Chilodon, 8.
Chilostomella, 6.
Chimæra, 90.
Chinchilla, 114.
Chirodota, 31.
Chirotes, 96.
Chiroteuthis, 84.
Chiton, 81.
Chlamyphorus, 108.
Chloëon, 68.
Cholœpus, 108.
Chondracanthus, 48.
Chondrilla, 9.
Chromis, 87.
Chrysochloris, 116.
Chrysopa, 69.
Chthamalus, 49.
Cicada, 66.
Cicindela, 72.
Ciconia, 101.
Cidaris, 27.
Ciona, 75.
Cirratulus, 46.
Cistudo, 94.
Clava, 16.
Clavellina, 75.
Cleodora, 83.
Clepsine, 45.
Climacograptus, 20.
Clio, 83.
Cliona, 10.
Clotho, 96.
Clupea, 87.
Clypeaster, 28.
Clytia, 17.
Coccinella, 71.
Coccocrinus, 32.
Coccus, 66.
Coccygus, 102.
Codonella, 8.
Codosiga, 8.
Cœlacanthus, 89.
Cœlopleurus, 27.
Cœnobita, 59.
Colaptes, 102.

INDEX OF GENERA.

Colias, 70.
Colius, 102.
Collocalia, 103.
Collozoum, 8.
*Collyrites, 28.
Colossendeus, 59.
Colpoda, 8.
Coluber, 95.
Columba, 102.
*Columnaria, 22.
Colymbus, 100.
Comatula, 32.
Condylura, 116.
Conochilus, 42.
Conus, 81.
Convoluta, 39.
Corallistes, 10.
Corallium, 23.
Corbis, 79.
Corbula, 79.
Corethra, 69.
Corixa, 66.
Coronula, 49.
Corophium, 55.
Corticium, 9.
Corvus, 102.
Corydalis, 69.
Coryne, 16.
*Coryphodon, 110.
Corystes, 59.
Cottus, 87.
Coturnix, 101.
Crabro, 71.
Crambessa, 20.
Cranchia, 84.
Crangon, 58.
Crania, 77.
Crassatella, 79.
Crax, 101.
Cribella, 29.
Criodrilus, 46.
Crisia, 74.
Cristatella, 74.
Cristellaria, 6.
Crocodilus, 98.
*Crotalocrinus, 32.
Crotalus, 96.
Cryptobranchus, 92.
*Cryptocrinus, 35.
Cryptoniscus, 57.
Cryptophialus, 49.
Ctenodiscus, 29.
Ctenodus, 91.
Ctenomys, 114.
Cucullanus, 41.
Cuculus, 102.
Cucumaria, 32.
Culcita, 29.
Culex, 69.
Cultellus, 79.
Cuma, 58.
*Cupressocrinus, 32.
Cyamus, 55.
Cyanea, 20.

*Cyathaxonia, 25.
*Cyathidium, 32.
*Cyathocrinus, 32.
*Cyathophyllum, 25.
Cyclas, 79.
Cyclodus, 97.
Cyclopina, 51.
Cyclops, 51.
Cyclostoma, 82.
Cyclothurus, 108.
Cygnus, 100.
Cymbulia, 83.
Cymothoa, 56.
Cynips, 71.
Cynocephalus, 118.
*Cynodraco, 98.
Cynthia, 75.
Cyprœa, 81.
Cypridina, 50.
Cyprina, 79.
Cyprinus, 87.
Cypris, 50.
Cypselus, 103.
Cyrena, 79.
*Cyrtoceras, 84.
*Cystiphyllum, 25.
Cystophora, 113.
Cythere, 50.
Cytherella, 51.

Dacelo, 103.
Dactylethra, 94.
Dactylocalyx, 10.
Dama, 110.
Dapedius, 88.
Daphnella, 53.
Daphnia, 53.
*Dasmia, 23.
Dasyprocta, 114.
Dasypus, 108.
Dasyurus, 107.
Degeeria, 66.
*Deinotherium, 112.
Delphinus, 109.
Demodex, 61.
Dendrobates, 94.
*Dendrocrinus, 32.
Dendrolagus, 107.
Dendrophis, 95.
Dendrophyllia, 22.
Dentalium, 80.
Depastrum, 20.
Diadema, 27.
Diastopora, 74.
Diastylis, 58.
*Diceras, 79.
*Dichograptus, 20.
Dicotyles, 110.
Dictyocha, 8.
*Dicynodon, 98.
Didelphis, 107.
Didemnum, 76.
Didunculus, 102.
*Didus, 102.

INDEX OF GENERA.

*Didymograptus, 20.
Didymophyes, 3.
Difflugia, 4.
*Dinoceras, 111.
*Dinornis, 99.
Diocus, 46.
Diodon, 88.
Diomedea, 100.
Diphasia, 17.
Diphyes, 19.
*Diplograptus, 20.
Diploria, 22.
Diplostoma, 37.
Diplozoön, 37.
Dipsas, 95.
*Dipterus, 91.
Dipus, 115.
Discina, 77.
Discoderma, 10.
Discoglossus, 94.
Discoporella, 74.
Discorbina, 6.
Distichopora, 21.
Distoma, 37.
Dochmius, 41.
Docophorus, 65.
Doliolum, 76.
Donax, 79.
Doris, 81.
Dorylaimus, 41.
Doto, 81.
Draco, 97.
Drassus, 64.
Dreissena, 78.
Dromaius, 99.
Dysidea, 9.

Echidna, 106.
Echinarachinus, 28.
Echinobothrium, 36.
Echinocyamus, 28.
Echinoderes, 42.
*Echinoencrinus, 35.
Echinolampas, 28.
Echinoneus, 28.
Echinorhynchus, 40.
*Echinothuria, 27.
Echinus, 26.
Echiurus, 43.
Edwardsia, 22.
Elaps, 96.
Eledone, 84.
Elephas, 112.
Emballonura, 115.
Emberiza, 102.
Embia, 68.
Emydium, 60.
Emys, 94.
Enchelys, 8.
Enchytræus, 46.
*Encrinus, 32.
*Endothyra, 5.
*Entomis, 50.
Eosphora, 42.

Epeira, 64.
Ephemera, 68.
Epicrium, 92.
Epistylis, 8.
Equus, 110.
Ericulus, 116.
Erinaceus, 116.
Errina, 21.
Erycina, 70.
*Eryon, 58.
Eschara, 74.
Esox, 87.
Estheria, 53.
*Eucalyptocrinus, 32.
Euchlanis, 42.
Eudendrium, 16.
*Eugeniacrinus, 32.
Euglena, 8.
Eunectes, 95.
Eunice, 46.
Euphausia, 57.
*Euphoberia, 64.
Euplectella, 10.
Euplotes, 8.
Euryale, 30.
Euspongia, 9, 10.
Eustrongylus, 41.

Falco, 104.
*Favosites, 22.
Felis, 113.
Fibularia, 28.
Filaria, 41.
Filograna, 46.
Firola, 82.
Fissurella, 81.
Flabellum, 23.
Floscularia, 42.
Flustra, 74.
Forficula, 67.
Formica, 71.
Fratercula, 100.
Fulgora, 66.
Fuligula, 100.
Fungia, 22.
*Fusulina, 6.
Fusus, 80.

Gadus, 87.
Galago, 117.
Galathea, 59.
Galeodes, 62.
Galeopithecus, 116.
*Galerites, 28.
Galeus, 90.
Gallinula, 101.
Gallus, 101.
Gamasus, 61.
Gammarus, 55.
Garrulus, 102.
*Gasterocoma, 32.
Gastrochæna, 79.
Gavialis, 98.
Gecarcinus, 59.

INDEX OF GENERA.

Gecko, 97.
Gemellaria, 74.
Geodia, 10.
Geometra, 70.
Geomys, 114.
Geophilus, 64.
Geoplana, 39.
Gerris, 66.
Globigerina, 6.
Glomeris, 64.
Glycera, 46.
*Glyptocrinus, 32.
*Glyptolœmus, 89.
*Glyptodon, 108.
*Glyptosphorites, 35.
Gobius, 87.
Goniaster, 29.
*Goniatites, 84.
*Goniophyllum, 25.
Gonodactylus, 57.
Gonoplax, 59.
Gordius, 40.
Gorgonia, 23.
Goura, 102.
Gracula, 102.
*Granatocrinus, 35.
Grantia, 10.
Grapsus, 59.
Gregarina, 3.
Gromia, 5.
Grus, 101.
Gryllotalpa, 67.
Gryllus, 67.
Gymnotus, 87.
Gynœcophorus, 37.
Gypaëtus, 104.
Gypogeranus, 104.
Gyrodactylus, 37.

*Habrocrinus, 32.
Hæmopsis, 45.
Halichœrus, 113.
Halichondria, 10.
Halicore, 109.
Haliotis, 81.
Halisarca, 9, 10.
*Halitherium, 109.
Halocypris, 50.
Halosydna, 46.
Halys, 96.
*Halysites, 23.
Hapale, 117.
*Haplocrinus, 32.
Harpacticus, 51.
Hatteria, 97.
Helarctos, 113.
Heliaster, 29.
Helicina, 82.
*Heliophyllum, 25.
Heliopora, 23.
Helix, 82.
Hemicardium, 79.
*Hemicidaris, 27.

*Hemicosmites, 35.
Hermella, 46.
Herpestes, 113.
Hesperia, 70.
*Hesperornis, 105.
Heterodesmus, 50.
Heterophrys, 8.
Heteropora, 74.
Hippa, 59.
*Hipparion, 110.
Hippobosca, 69.
Hippocampus, 88.
Hippolyte, 58.
Hippopotamus, 110.
Hippopus, 79.
Hippothoa, 74.
*Hippurites, 79.
Hirudo, 45.
Hirundo, 103.
Histriobdella, 44.
*Holocystis, 25.
*Holoptychius, 89.
Holopus, 32.
Holothuria, 32./.
Holtenia, 10.
Homarus, 58.
Homo, 118.
Horniphora, 25.
Hornera, 74.
Hyæna, 113.
Hyalea, 83.
Hyalonema, 10.
Hyalosphenia, 4.
Hyas, 59.
*Hybodus, 90.
Hydatina, 42.
Hydra, 14.
Hydrachna, 61.
Hydractinia, 16.
Hydrochœrus, 114.
Hydrometra, 66.
Hydrophis, 96.
Hyla, 94.
Hylobates, 118.
Hymenaster, 30.
Hymeniacidon, 10.
*Hyolithes, 83.
Hyomoschus, 111.
Hyperia, 55.
Hyperoödon, 109.
Hypsiprymnus, 107.
Hyrax, 112.
Hystrix, 114.

Ibis, 101.
Ichneumon, 71.
Ichthydium, 42.
*Ichthyornis, 105.
*Ichthyosaurus, 98.
Idmonea, 74.
Idotea, 56.
Idyia, 25.
Iguana, 97.

INDEX OF GENERA.

*Iguanodon, 98.
*Illanus, 54.
Ilyanthus, 22.
Inachus, 59.
*Inoceramus, 78.
Indris, 117.
Isis, 23.
Isocardia, 79.
Isodictya, 10.
Iulus, 64.
Ixodes, 61.

*Koninckina, 77.
Kröyera, 55.

Labrus, 87.
*Labyrinthodon, 94.
Lacerta, 97.
Lafoëa, 17.
Lagena, 6.
Lagomys, 114.
Lagopus, 101.
Lamna, 90.
Lanius, 103.
Larus, 100.
Lecanium, 66.
Leda, 78.
*Leiodon, 97.
Lemur, 117.
Lepas, 49.
Lepidonotus, 46.
Lepidosiren, 91.
Lepidosteus, 88.
*Lepidotus, 88.
Lepidurus, 53.
Lepisma, 66.
Lepralia, 74.
*Leptolepis, 88.
Leptoplana, 39.
Lepus, 114.
Lernæa, 48.
Lernæodiscus, 48.
Leucifer, 57.
Leucosia, 59.
Leucosolenia, 10.
Libellula, 68.
Ligia, 57.
Ligula, 36.
Lima, 78.
Limacina, 83.
Limax, 82.
Limnadia, 53.
Limnæa, 82.
Limnochares, 61.
Limnocythere, 50.
Limnodrilus, 46.
Limnophilus, 69.
Limnoria, 56.
Limulus, 54.
Linckia, 29.
Lineus, 39.
Lingula, 77.
Lithobius, 64.

Lithodes, 59.
Littorina, 81.
*Lituites, 84.
Lituola, 5.
Locusta, 67.
*Loftusia, 5.
Loligo, 84.
Loligopsis, 84.
Lophius, 87.
Lophogaster, 57.
Lophohelia, 23.
Lophopus, 74.
Lorius, 102.
*Loxomma, 94.
Loxosoma, 75.
Lucernaria, 20.
Lucina, 79.
Luffaria, 9.
Luidia, 29.
Lumbriconereis, 46.
Lumbricus, 46.
Lutra, 113.
Lutraria, 79.
Lycæna, 70.
Lychnocanium, 8.
Lycosa, 63.
Lynceus, 53.

Macacus, 118.
MacAndrewia, 10.
*Machairodus, 113.
Machilis, 66.
*Macrauchenia, 110.
Macrobiotus, 60.
Macrocercus, 102.
Macropus, 107.
Macrosceles, 117.
Macrostomum, 39.
Mactra, 79.
Madrepora, 22.
Maia, 59.
Malacobdella, 44.
Malleus, 78.
*Malocystites, 35.
Manatus, 109.
Manis, 108.
Mantis, 67.
Marginulina, 6.
*Marsupites, 32.
Mastigamœba, 4.
*Mastodon, 112.
*Mastodonsaurus, 94.
Meandrina, 22.
*Megalonyx, 108.
*Megalosaurus, 98.
Megalotrocha, 42.
Megapodius, 101.
*Megatherium, 108.
Melania, 81.
Meleagris, 101.
Meles, 113.
Melicerta, 42.
Meliphaga, 103.

Mellita, 28.
Mellivora, 113.
*Melocrinus, 32.
Melophagus, 69.
Membranipora, 74.
Menobranchus, 92.
Menopoma, 92.
Mermis, 40.
Merops, 103.
Merulina, 22.
*Mesotherium, 115.
*Michelinia, 22.
*Micraster, 28.
Microgromia, 5.
Midas, 117.
Miliola, 5.
Millepora, 21.
Milvus, 104.
Minyas, 21.
Mitra, 81.
Molgula, 75.
Molpadia, 32.
Monas, 8.
Monocystis, 3.
Monodon, 109.
*Monograptus, 20.
Monosiga, 8.
Monostomum, 37.
*Monticulipora, 25.
Mopsea, 23.
*Mosasaurus, 97.
Motacilla, 103.
Mugil, 87.
Munna, 56.
Munnopsis, 56.
Murœna, 87.
Murex, 80.
Mus, 114.
Musca, 69.
Muscicapa, 103.
Musophaga, 102.
Mustela, 113.
Mya, 79.
Mycetes, 117.
Mygale, 63.
Myliobatis, 90.
*Mylodon, 108.
Myodes, 114.
Myogale, 117.
Myopotamus, 114.
Myoxus, 115.
Myrmecophaga, 108.
Myrmeleo, 69.
Mysis, 57.
Mytilus, 78.
Myxastrum, 4.
Myxine, 87.
Myxodictyon, 4.
Myzostoma, 37.

Nais, 46.
Naja, 96.
Nassa, 81.
Nasua, 113.
Natica, 81.
Nautilus, 84.
Nebalia, 53, 54.
Nectarinia, 103.
Nemertes, 39.
Neophron, 104.
Nepa, 66.
Nephelis, 45.
Nephrops, 58.
Nephthys, 46.
Nereis, 46.
Nerine, 46.
Nerita, 81.
Nestor, 102.
Nicothoe, 48.
Noctua, 70.
Nodosaria, 6.
Notidanus, 90.
Notodelphys (*Crustacea*), 51 ; (*Amphibia*), 94.
Notommata, 42.
Notonecta, 66.
Nubecularia, 5.
*Nucleocrinus, 35.
Nucleolites, 28.
Nucula, 78.
Nuculana, 78.
Numenius, 101.
Nummulites, 6.
Nycteribia, 69.
Nycticebus, 117.
Nyctotherus, 8.
Nymphalis, 70.
Nymphon, 59.

Obelia, 17.
Obisium, 62.
Octodon, 114.
Octopus, 84.
Oculina, 23.
Ocypoda, 59.
Œcistes, 42.
Œdipoda, 67.
Œstrus, 69.
Oliva, 81.
Ommastrephes, 84.
Oncinolabes, 31.
Oniscus, 57.
Ophioconna, 30.
Ophioglypha, 30.
Ophiolepis, 30.
Ophiophagus, 96.
Ophiura, 30.
Opilio, 61.
Opisthocomus, 101.
Opisthomum, 39.
*Orbitoides, 6.
Orbitolites, 5.
Orchestia, 55.
*Oreodon, 110.
Oribates, 61.
Oriolus, 102.

INDEX OF GENERA.

Ornithorhynchus, 106.
*Orohippus, 110.
Orthagoriscus, 88.
*Orthis, 77.
*Orthoceras, 84.
Orycteropus, 108.
Osculina, 9.
*Osteolepis, 89.
Ostracion, 88.
Ostrea, 78.
Otaria, 113.
Otis, 101.
*Oudenodon, 98.
Ovibos, 111.
Ovis, 111.
Ovulum, 81.
Oxyuris, 41.

Pagurus, 59.
*Palæaster, 30.
*Palæchinus, 27.
Palæmon, 58.
*Palæoniscus, 88.
*Palæotherium, 110.
Palamedea, 101.
Palinurus, 58.
Pallene, 60.
Palmipes, 29.
Paludina, 81.
Palythoa, 22.
Pandalus, 58.
Panorpa, 69.
Papilio, 70.
Paradisea, 102.
*Paradoxides, 54.
Paramœcium, 8.
*Parkeria, 5.
Parra, 101.
Parus, 103.
Passer, 102.
Patella, 81.
Pauropus, 64.
Pavo, 101.
Pavonaria, 23.
Pecten, 78.
Pectinaria, 46.
Pectunculus, 78.
Pedalion, 42.
Pedicellina, 75.
Pediculus, 65.
Pelagia, 20.
Pelagonemertes, 39.
Pelecanus, 100.
Pelias, 96.
Pelobates, 94.
Pelomyxa, 4.
Pelonaia, 75.
*Peltocaris, 53.
Peltogaster, 48.
Penæus, 58.
Peneroplis, 5.
Pennatula, 23.
Pentacrinus, 32, 34.

Pentacta, 32.
*Pentamerus, 77.
Pentastoma, 60.
Pentatoma, 67.
*Pentremites, 35.
Perameles, 107.
Perca, 87.
Perdix, 101.
Peridinium, 8.
*Periechocrinus, 32.
Peripatus, 64.
Periplaneta, 67.
Perla, 68.
Perophora, 75.
Petaurus, 107.
Petromyzon, 87.
Phacochœrus, 110.
Phaeton, 100.
Phalacrocorax, 100.
Phalangista, 107.
Phalangium, 61.
*Phaneropleuron, 89.
Phanogenia, 32.
Phascolarctos, 107.
Phascolomys, 106.
Phascolosoma, 43.
Phasianus, 101.
Phasma, 67.
Philodina, 42.
Phoca, 113.
Phocæna, 109.
Phœnicopterus, 100.
*Pholadomya, 79.
Pholas, 79.
Phormosoma, 27.
Phoronis, 43.
Phoxichilus, 60.
Phronima, 55.
Phryganea, 69.
Phrynus, 62.
Phthirius, 65.
Phyllium, 67.
Phyllobothrium, 36.
Phyllodoce, 46.
*Phyllograptus, 20.
Phyllomedusa, 94.
Phyllostoma, 115.
Phylloxera, 66.
Physalia, 20.
Physeter, 109.
Physophora, 20.
Pica, 102.
Picus, 102.
Pilumnus, 59.
Pinna, 78.
Pinnotheres, 59.
Pipa, 94.
Piscicola, 45.
*Pisocrinus, 32.
Placuna, 78.
Planaria, 39.
Platalea, 101.
Platanista, 109.

INDEX OF GENERA.

Platurus, 96.
*Platycrinus, 32.
*Platysomus, 88.
Plecotus, 115.
*Plesiosaurus, 98.
Plethodon, 92.
Pleurobrachia, 25.
Pleuronectes, 87.
Pleurotoma, 81.
Pleurotomaria, 81.
*Plicatocrinus, 32.
Pliobothrus, 21.
*Pliosaurus, 98.
Plotus, 100.
Plumatella, 74.
Plumularia, 17.
Plyctolophus, 102.
Pneumodermon, 83.
Podiceps, 100.
Podocyrtis, 8.
Podophrya, 8.
Podura, 66.
Pœcilasma, 49.
Pollicipes, 49.
Polyarthra, 42.
Polycelis, 39.
Polyclonia, 20.
Polycope, 50.
Polydesmus, 64.
Polyergus, 71.
Polygordius, 47.
Polynoë, 46.
Polyodon, 89.
Polyphemus, 53.
Polypterus, 89.
Polystomum, 37.
Polyxenus, 64.
Polyzonium, 64.
Pomacentrus, 87.
Ponera, 71.
Pontella, 51.
Pontobdella, 45.
Porcellana, 59.
Porites, 22.
Porocidaris, 27.
Porpita, 20.
Portunus, 59.
Potamogale, 116.
*Poteriocrinus, 32.
Praya, 18, 19.
Priapulus, 43.
Pristiophorus, 90.
Pristis, 90.
Procellaria, 100.
Procyon, 113.
*Producta, 77.
Promerops, 103.
Prostomum, 39.
Protamœba, 4.
Protella, 55.
Proteolepas, 49.
Proteus, 92.
Protomyxa, 4.

Protopterus, 91.
*Protorosaurus, 97.
Psammobia, 79.
Pseudis, 94.
Pseudopus, 97.
Psittacus, 102.
Psocus, 68.
Psolus, 32.
*Pteranodon, 98.
Pteraster, 30.
*Pterichthys, 89.
Pteroceras, 80.
Pterocles, 101.
*Pterodactylus, 98.
Pteropus, 116.
*Pterygotus, 54.
Pulex, 69.
Pulvinulina, 6.
Pupa, 82.
Purpura, 81.
Putorius, 113.
*Pycnodus, 88.
Pycnogonum, 60.
Pygaster, 28.
Pyramidella, 81.
Pyrgoma, 49.
Pyrosoma, 76.
Python, 95.
Pyxis, 94.

Quadrula, 4.

Raia, 90.
Rallus, 101.
Rana, 94.
Ranina, 59.
Reduvius, 66.
Reniera, 10.
Renilla, 23.
Retepora, 74.
*Retiolites, 20.
Rhabditis, 41.
Rhabdogaster, 41.
Rhabdopleura, 75.
Rhamphastos, 102.
*Rhamphorhynchus, 98.
Rhaphigaster, 67.
Rhea, 99.
Rhesus, 118.
Rhina, 90.
Rhinobatis, 90.
Rhinoceros, 110.
Rhinodon, 90.
Rhinolophus, 115.
Rhinophrynus, 94.
Rhipidogorgia, 23, 24.
Rhizostoma, 20.
Rhizoxenia, 23.
*Rhodocrinus, 32.
Rhombus, 87.
Rhynchonella, 77.
Rhynchopygus, 28.
*Rhytina, 109.

INDEX OF GENERA.

Rossia, 84.
Rotalia, 6.
Rotifer, 42.
Rotula, 28.
Runcina, 81.
Rupicapra, 111.

Sabella, 46.
Sabellaria, 46.
Saccammina, 5.
Sacculina, 43.
Sænuris, 46.
Sagartia, 21.
Sagitta, 47.
Salamandra, 92.
Salenia, 27.
Salmo, 87.
Salpa, 76.
Salticus, 63.
Sanguisuga, 45.
Sarcopsylla, 69.
Sarcoptes, 61.
Sarcorhamphus, 104.
Saxicava, 79.
Saxicola, 103.
Scalaria, 81.
Scalpellum, 49.
Scaphirhynchus, 89.
Scaphites, 84.
Scincus, 97.
Sciurus, 115.
Sclerostoma, 41.
Scolopax, 101.
Scolopendra, 64.
Scomber, 87.
Scops, 104.
Scorpio, 62.
Scrobicularia, 79.
Scruparia, 74.
Scutella, 28.
Scutellera, 67.
Scutigera, 64.
Scyllarus, 58.
Scyllium, 90.
Seison, 42.
Selache, 90.
Selenaria, 74.
Semnopithecus, 118.
Sepia, 84.
Sepiola, 84.
Seps, 97.
Serolis, 56.
Serpula, 46.
Sertularia, 17.
Sida, 53.
Sigalion, 46.
Sigaretus, 81.
Silurus, 87.
Simia, 118.
*Siphonia, 10.
Siphonops, 92.
Sipunculus, 43.
Siredon, 92.

Siren, 92.
Sirex, 71.
Sitta, 103.
*Slimonia, 54.
Smynthurus, 66.
Solanocrinus, 32.
Solarium, 81.
Solaster, 29.
Solea, 87.
Solen, 79.
Solenostoma, 88.
Sorex, 116.
Spalax, 114.
Spatangus, 28.
Spatularia, 89.
Sphæroma, 56.
*Sphæronites, 35.
Sphærozoum, 8.
Sphærularia, 40.
Sphargis, 94.
Spheniscus, 99.
Sphinx, 70.
Spinax, 90.
Spio, 46.
Spirialis, 83.
*Spirifera, 77.
Spiroptera, 41.
Spirorbis, 46.
Spirula, 84.
Spondylus, 78.
Spongilla, 10.
*Squalodon, 109.
Squilla, 57.
*Stagonolepis, 98.
*Stauria, 25.
Steatornis, 103.
Stellio, 97.
Stenops, 117.
Stenorhynchus, 59.
Stentor, 8.
Stephanoceros, 42.
Stephanomia, 20.
Stephanoscyphus, 17.
Sterna, 100.
Sternaspis, 43.
Stomoxys, 69.
*Streptospondylus, 98.
Strigops, 102.
Strix, 104.
Strombus, 80.
*Strophomena, 76.
Struthio, 99.
Sturnus, 102.
Stylaster, 21.
Stylops, 71.
Stylorhynchus, 3.
Sula, 100.
Sulcator, 55.
Sus, 110.
Syllis, 46.
Sylvia, 103.
Symbranchus, 87.
Synapta, 31.

I

Syngamus, 88.
*Syringopora, 22.
Syrphus, 69.

Tabanus, 69.
Tachypetes, 100.
Tænia, 36.
Talitrus, 55.
Talpa, 116.
Tanagra, 102.
Tanais, 56.
Tantalus, 101.
Tapirus, 110.
Tarsius, 117.
*Taxocrinus, 32.
Tealia, 21.
Tegenaria, 64.
Tejus, 97.
*Teleosaurus, 98.
Tellina, 79.
Telphusa, 59.
Temnechinus, 27.
Tenthredo, 71.
Terebella, 46.
Terebratula, 77.
Teredo, 79.
Termes, 68.
Testudo, 94.
Tethya, 10.
Tethys, 81.
*Tetradium, 23.
*Tetragraptus, 20.
Tetranychus, 61.
Tetrao, 101.
Tetrarhynchus, 36.
Tetrastemma, 39.
Textularia, 6.
Thalassema, 43.
Thalassicolla, 8.
Thalassidroma, 100.
Thalassina, 58.
Thalassolampe, 8.
Thaumantias, 17.
*Theca, 83.
*Thecia, 23.
Thecidium, 77.
Thecla, 70.
Thelyphonus, 62.
Theridium, 64.
Thomisus, 64.
Thracia, 79.
Thrips, 67.
Thylacinus, 107.
Thyone, 32.
*Tillotherium, 111.
Tinamus, 101.
Tinea, 70.
Tipula, 69.
Tomopteris, 46, 47.
Tornatella, 81.
Torpedo, 90.
Tortrix (*Insecta*), 70; (*Reptilia*), 95.
*Toxodon, 112.

Tracheliaster, 48.
Trachynema, 17, 18.
Tragulus, 111.
Treron, 102.
*Triacrinus, 32.
Trichecus, 113.
Trichina, 41.
Trichocephalus, 41.
Trichodectes, 65.
Trichoderia, 8.
Trichoglossus, 102.
Tridacna, 79.
Trigonia, 78.
*Trimerella, 77.
Trimeresurus, 96.
Trionyx, 94.
*Tristichopterus, 89.
Tristoma, 37.
Triton, 92.
Trochammina, 5.
Trochetia, 45.
Trochilus, 103.
Trochus, 81.
Troglodytes (*Aves*), 103; (*Mammalia*), 118.
Trogon, 102.
Tropidonotus, 95.
Trygon, 90.
Tubifex, 46.
Tubipora, 23.
Tubularia, 16.
Tubulipora, 74.
Tupaia, 116.
Turbinolia, 23.
Turbo, 81.
Turdus, 103.
Turnix, 101.
Turritella, 81.
Turtur, 102.
Tylenchus, 41. *Tylopoda*, 111.
Typhlops, 95.
Tyrannus, 103.

Umbrella, 81.
Unio, 78.
Upupa, 103.
Uraster, 29.
Uria, 100.
Uropeltis, 95.
Ursus, 113.

Valkeria, 74.
Vanellus, 101.
Vanessa, 70.
Varanus, 97.
Velella, 20.
*Ventriculites, 10.
Venus, 79.
Veretillum, 23.
Vermetus, 81.
Verruca, 49.
Vesicularia, 74.
Vespa, 71.

INDEX OF GENERA.

Vespertilio, 115.
Vincularia, 74.
Vipera, 96.
Virgularia, 23.
Viverra, 113.
Vogtia, 19.
Volucella, 69.
Voluta, 81.
Vorticella, 8.
Vulpes, 113.
Vultur, 104.

Xiphacantha, 7.

Yunx, 102.

**Zaphrentis,* 25.
**Zeacrinus,* 32.
**Zeuglodon,* 109.
Ziphius, 109.
Zoanthus, 22.
Zonurus, 97.
Zootoca, 97.

THE END.

WORKS BY THE SAME AUTHOR.

I.
A MANUAL OF ZOOLOGY.
FOR THE USE OF STUDENTS.
WITH A GENERAL INTRODUCTION ON THE PRINCIPLES OF ZOOLOGY.

Sixth Edition, Revised and greatly Enlarged.
Crown 8vo, pp. 865, with 394 Engravings on Wood. 14s.

"It is the best manual of zoology yet published, not merely in England, but in Europe." —*Pall Mall Gazette.*

"We hold that it would be difficult indeed to find a work which gives, in so brief a compass, so luminous and philosophical a view of the whole Animal Kingdom. To the earnest student entering upon the science of Biology, the 'General Introduction' alone must be a boon of the highest order."—*Quarterly Journal of Science.*

"As a general systematic treatise on the structure and classification of Animals, it is the best which we possess."—*Annals and Magazine of Natural History.*

II.
TEXT-BOOK OF ZOOLOGY.
FOR THE USE OF SCHOOLS.

Third Edition, Enlarged. Crown 8vo, with 188 Engravings on Wood. 6s.

"This capital introduction to natural history is illustrated and well got up in every way. We should be glad to see it generally used in schools."—*Medical Press and Circular.*

III.
INTRODUCTORY TEXT-BOOK OF ZOOLOGY.
FOR THE USE OF JUNIOR CLASSES.

Fifth Edition, Revised and Enlarged, with 156 Engravings. 3s.

"Very suitable for junior classes in schools. There is no reason why any one should not become acquainted with the principles of the science, and the facts on which they are based, as set forth in this volume."—*Lancet.*

"Nothing can be better adapted to its object than this cheap and well-written Introduction."—*London Quarterly Review.*

IV.
OUTLINES OF NATURAL HISTORY.
FOR BEGINNERS.
BEING DESCRIPTIONS OF A PROGRESSIVE SERIES OF ZOOLOGICAL TYPES.

Second Edition. With 52 Engravings. 1s. 6d.

"There has been no book since Patterson's well-known 'Zoology for Schools' that has so completely provided for the class to which it is addressed as the capital little volume by Dr Nicholson."—*Popular Science Review.*

V.
INTRODUCTION TO THE STUDY OF BIOLOGY.

Crown 8vo, with numerous Engravings. 5s.

BY THE SAME AUTHOR—continued.

VI.

EXAMINATIONS IN NATURAL HISTORY.

BEING A PROGRESSIVE SERIES OF QUESTIONS ADAPTED TO THE AUTHOR'S INTRODUCTORY AND ADVANCED TEXT-BOOKS AND THE STUDENT'S MANUAL OF ZOOLOGY.

Crown 8vo, 1s.

VII.

ON THE STRUCTURE AND AFFINITIES OF

THE "TABULATE CORALS" OF THE PALÆOZOIC PERIOD.

WITH CRITICAL DESCRIPTIONS OF ILLUSTRATIVE SPECIES.

Illustrated with Engravings on Wood, and 15 Lithograph Plates. Royal 8vo, 21s.

VIII.

A MANUAL OF PALÆONTOLOGY.

FOR THE USE OF STUDENTS.

WITH A GENERAL INTRODUCTION ON THE PRINCIPLES OF PALÆONTOLOGY.

Second Edition. 2 Vols. crown 8vo, with 722 Engravings. 42s.

"This book will be found to be one of the best of guides to the principles of Palæontology and the study of organic remains."—*Athenæum.*
"The woodcuts are simply perfection."—*Graphic.*

IX.

THE ANCIENT LIFE-HISTORY OF THE EARTH.

AN OUTLINE OF THE PRINCIPLES AND LEADING FACTS OF PALÆONTOLOGICAL SCIENCE. WITH A GLOSSARY AND INDEX.

Crown 8vo, with 270 Engravings. 10s. 6d.

"By a master in the science, who understands the significance of every phenomenon which he records, and knows how to make it reveal its lessons. As regards the value of the work, there can scarcely exist two opinions. As a text-book of the historical phase of Palæontology it will be indispensable to students, whether specially pursuing Geology or Biology; and without it no man who aspires even to an outline knowledge of Natural Science can deem his library complete."—*Quarterly Journal of Science.*

X.

MONOGRAPH OF THE BRITISH GRAPTOLITIDÆ.

PART I.—GENERAL INTRODUCTION.

With 74 Engravings, 8vo, pp. 133. 5s.

WILLIAM BLACKWOOD & SONS, EDINBURGH AND LONDON.

Works on Mental Philosophy.

LECTURES ON METAPHYSICS. By Sir WILLIAM HAMILTON, Bart., Professor of Logic and Metaphysics in the University of Edinburgh. Edited by the Very Rev. H. L. MANSELL, LL.D., Dean of St Paul's, and JOHN VEITCH, M.A., Professor of Logic and Rhetoric, Glasgow. Sixth Edition. 2 vols. 8vo. 24s.

LECTURES ON LOGIC. By Sir WILLIAM HAMILTON, Bart. Edited by the Same. Third Edition. 2 vols. 8vo. 24s.

DISCUSSIONS ON PHILOSOPHY AND LITERATURE, EDUCATION AND UNIVERSITY REFORM. By SIR WILLIAM HAMILTON, Bart. Third Edition. 8vo. 21s.

PHILOSOPHICAL WORKS OF THE LATE JAMES FREDERICK FERRIER, B.A. Oxon., LL.D., Professor of Moral Philosophy and Political Economy in the University of St Andrews. New Edition. 3 vols. crown 8vo. 34s. 6d.

The following are sold Separately:—

INSTITUTES OF METAPHYSIC. Third Edition. 10s. 6d.

LECTURES ON THE EARLY GREEK PHILOSOPHY. Second Edition. 10s. 6d.

PHILOSOPHICAL REMAINS, INCLUDING THE LECTURES ON EARLY GREEK PHILOSOPHY. Edited by Sir ALEX. GRANT, Bart., D.C.L., and Professor LUSHINGTON. 2 vols. 24s.

PORT ROYAL LOGIC. TRANSLATED FROM THE FRENCH: with Introduction, Notes, and Appendix. By THOMAS SPENCER BAYNES, LL.D., Professor of Logic and English Literature in the University of St Andrews. Eighth Edition, 12mo. 4s.

METHOD, MEDITATIONS, AND PRINCIPLES OF PHILOSOPHY OF DESCARTES. Translated from the original French and Latin. With a New Introductory Essay, Historical and Critical, on the Cartesian Philosophy. By JOHN VEITCH, LL.D., Professor of Logic and Rhetoric in the University of Glasgow. Eighth Edition, 12mo. 6s. 6d.

WILLIAM BLACKWOOD & SONS, EDINBURGH AND LONDON.

WILLIAM BLACKWOOD & SONS'
RECENT PUBLICATIONS.

KOUMISS; OR, FERMENTED MARE'S MILK: AND ITS USES IN THE TREATMENT AND CURE OF PULMONARY CONSUMPTION AND OTHER WASTING DISEASES. With an Appendix on the best Methods of Fermenting Cow's Milk. By GEORGE L. CARRICK, M.D., L.R.C.S.E. and L.R.C.P.E., Physician to the British Embassy, St Petersburg, &c. Crown 8vo, 10s. 6d.

"Nor is Dr Carrick content with mere assertions, however much they may commend themselves by their reasonableness, but cites some remarkable facts in proof of his contention.......The statistics of Dr Carrick will startle those most sceptical of the value of Koumiss as a remedy for consumptive diseases."—*Athenæum*.

"Let every one who is interested in a consumptive patient read the book and judge for himself.......Dr Carrick gives a full account both of the various methods of preparing Koumiss, and of the best means of preserving and employing it....... We may cordially praise his book, which will perhaps in many cases prove as useful as it is certainly interesting."—*Saturday Review*.

"Dr Carrick deals very thoroughly with the history of Koumiss—its constituents, its therapeutic action, and even the best way of reaching the somewhat remote localities where it can most beneficially be drunk.......His work is pleasantly and lucidly written."—*St James's Gazette*.

"Dr Carrick writes with judgment, and shows a complete knowledge of his subject. Physicians should all read his book."—*Edinburgh Courant*.

"Dr Carrick deserves credit for the complete manner in which he has conducted his investigations, and also for the clear and forcible language in which he states his views."—*Dundee Advertiser*.

ELEMENTARY HAND-BOOK OF PHYSICS. With 210 Diagrams. By WILLIAM ROSSITER, F.R.A.S., &c. Crown 8vo, pp. 390. 5s.

"A singularly interesting Treatise on Physics, founded on facts and phenomena gained at first hand by the Author, and expounded in a style which is a model of that simplicity and ease in writing which betokens mastery of the subject. To those who require a non-mathematical exposition of the principles of Physics, a better book cannot be recommended."—*Pall Mall Gazette*.

PROFESSOR JOHNSTON'S CHEMISTRY OF COMMON LIFE. New Edition, revised and brought down to the present time. By ARTHUR HERBERT CHURCH, M.A. Oxon., Author of 'Food, its Sources, Constituents, and Uses;' 'The Laboratory Guide for Agricultural Students,' &c. Illustrated with Maps and 102 Engravings on Wood. Crown 8vo, pp. 618. 7s. 6d.

"No popular scientific work that has ever been published has been more generally and deservedly appreciated than the late Professor Johnston's 'Chemistry of Common Life.'......It remains unrivalled as a clear, interesting, comprehensive, and exact treatise upon the important subjects with which it deals.......The book is one which not only every student but every educated person who lives should read, and keep to refer to."—*Mark Lane Express*.

A MANUAL OF BOTANY, ANATOMICAL AND PHYSIOLOGICAL. For the Use of Students. By ROBERT BROWN, M.A., PH.D., F.R.G.S. Crown 8vo, with numerous Illustrations. 12s. 6d.

"We have no hesitation in recommending this volume to our readers as being the best and most reliable of the many works on botany yet issued.......His manual will, if we mistake not, be eagerly consulted and attentively studied by all those who take an interest in the science of botany."—*Civil Service Gazette*.

WILLIAM BLACKWOOD & SONS, EDINBURGH AND LONDON.

CATALOGUE

OF

MESSRS BLACKWOOD & SONS' PUBLICATIONS.

PHILOSOPHICAL CLASSICS FOR ENGLISH READERS.

EDITED BY WILLIAM KNIGHT, LL.D.,
Professor of Moral Philosophy in the University of St Andrews.
In crown 8vo Volumes, with Portraits, price 3s. 6d.

Now ready—

I. **Descartes.** By Professor MAHAFFY, Dublin.
II. **Butler.** By Rev. W. LUCAS COLLINS, M.A.
III. **Berkeley.** By Professor FRASER, Edinburgh.
IV. **Fichte.** By Professor ADAMSON, Owens College, Manchester.
V. **Kant.** By WILLIAM WALLACE, M.A., LL.D., Merton College, Oxford.

The Volumes in preparation are—

HAMILTON. By Professor Veitch, Glasgow.
HUME. By the Editor.
BACON. By Professor Nichol, Glasgow.
HEGEL. By Professor Edward Caird, Glasgow.

HOBBES. By Professor Croom Robertson, London.
SPINOZA. By Dr Martineau, Principal of Manchester New College.
VICO. By Professor Flint, Edinburgh.

IN COURSE OF PUBLICATION.

FOREIGN CLASSICS FOR ENGLISH READERS.

EDITED BY MRS OLIPHANT.

In Crown 8vo, 2s. 6d.

The Volumes published are—

DANTE. By the Editor.
VOLTAIRE. By Major-General Sir E. B. Hamley, K.C.M.G.
PASCAL. By Principal Tulloch.
PETRARCH. By Henry Reeve, C.B.
GOETHE. By A. Hayward, Q.C.
MOLIÈRE. By the Editor and F. Tarver, M.A.
MONTAIGNE. By Rev. W. L. Collins, M.A.
RABELAIS. By Walter Besant, M.A.

CALDERON. By E. J. Hasell.
SAINT SIMON. By Clifton W. Collins, M.A.
CERVANTES. By the Editor.
CORNEILLE AND RACINE. By Henry M. Trollope.
MADAME DE SÉVIGNÉ. By Miss Thackeray.
LA FONTAINE, AND OTHER FRENCH FABULISTS. By Rev. W. Lucas Collins, M.A.
SCHILLER. By James Sime, M.A., Author of 'Lessing: his Life and Writings.'

In preparation—
ROUSSEAU. By Henry Graham.—TASSO. By E. J. Hasell.

NOW COMPLETE.

ANCIENT CLASSICS FOR ENGLISH READERS.

EDITED BY THE REV. W. LUCAS COLLINS, M.A.

Complete in 28 Vols. crown 8vo, cloth, price 2s. 6d. each. And may also be had in 14 Volumes, strongly and neatly bound, with calf or vellum back, £3, 10s.

Saturday Review.—"It is difficult to estimate too highly the value of such a series as this in giving 'English readers' an insight, exact as far as it goes, into those olden times which are so remote and yet to many of us so close."

CATALOGUE

OF

MESSRS BLACKWOOD & SONS'
PUBLICATIONS.

ALISON. History of Europe. By Sir ARCHIBALD ALISON, Bart., D.C.L.
1. From the Commencement of the French Revolution to the Battle of Waterloo.
   LIBRARY EDITION, 14 vols., with Portraits. Demy 8vo, £10, 10s.
   ANOTHER EDITION, in 20 vols. crown 8vo, £6.
   PEOPLE'S EDITION, 13 vols. crown 8vo, £2, 11s.
2. Continuation to the Accession of Louis Napoleon.
   LIBRARY EDITION, 8 vols. 8vo, £6, 7s. 6d.
   PEOPLE'S EDITION, 8 vols. crown 8vo, 34s.
3. Epitome of Alison's History of Europe. Twenty-ninth Thousand, 7s. 6d.
4. Atlas to Alison's History of Europe. By A. Keith Johnston.
   LIBRARY EDITION, demy 4to, £3, 3s.
   PEOPLE'S EDITION, 31s. 6d.

—— Some Account of my Life and Writings: an Autobiography of the late Sir Archibald Alison, Bart., D.C.L. Edited by his Daughter-in-law. 2 vols. 8vo, with Portrait engraved on Steel. [*In the Press.*

—— Life of John Duke of Marlborough. With some Account of his Contemporaries, and of the War of the Succession. Third Edition, 2 vols. 8vo. Portraits and Maps, 30s.

—— Essays: Historical, Political, and Miscellaneous. 3 vols. demy 8vo, 45s.

—— Lives of Lord Castlereagh and Sir Charles Stewart, Second and Third Marquesses of Londonderry. From the Original Papers of the Family. 3 vols. 8vo, £2, 2s.

—— Principles of the Criminal Law of Scotland. 8vo, 18s.

—— Practice of the Criminal Law of Scotland. 8vo, cloth boards, 18s.

—— The Principles of Population, and their Connection with Human Happiness. 2 vols. 8vo, 30s.

ALISON. On the Management of the Poor in Scotland, and its Effects on the Health of the Great Towns. By WILLIAM PULTENEY ALISON, M.D. Crown 8vo, 5s. 6d.

ADAMS. Great Campaigns. A Succinct Account of the Principal Military Operations which have taken place in Europe from 1796 to 1870. By Major C. ADAMS, Professor of Military History at the Staff College. Edited by Captain C. COOPER KING, R.M. Artillery, Instructor of Tactics, Royal Military College. 8vo, with Maps. 16s.

LIST OF BOOKS PUBLISHED BY

AIRD. Poetical Works of Thomas Aird. Fifth Edition, with Memoir of the Author by the Rev. JARDINE WALLACE, and Portrait. Crown 8vo, 7s. 6d.

—— The Old Bachelor in the Old Scottish Village. Fcap. 8vo, 4s.

ALLARDYCE. The City of Sunshine. By ALEXANDER ALLARDYCE. Three vols. post 8vo, £1, 5s. 6d.

—— Memoir of the Honourable George Keith Elphinstone, K.B., Viscount Keith of Stonehaven Marischal, Admiral of the Red. One vol. 8vo, with Portrait, Illustrations, and Maps. [*In the Press.*

ANCIENT CLASSICS FOR ENGLISH READERS. Edited by Rev. W. LUCAS COLLINS, M.A. Complete in 28 vols., cloth, 2s. 6d. each; or in 14 vols., tastefully bound, with calf or vellum back, £3, 10s.

Contents of the Series.

HOMER: THE ILIAD. By the Editor.
HOMER: THE ODYSSEY. By the Editor.
HERODOTUS. By George C. Swayne, M.A.
XENOPHON. By Sir Alexander Grant, Bart., LL.D.
EURIPIDES. By W. B. Donne.
ARISTOPHANES. By the Editor.
PLATO. By Clifton W. Collins, M.A.
LUCIAN. By the Editor.
ÆSCHYLUS. By the Right Rev. the Bishop of Colombo.
SOPHOCLES. By Clifton W. Collins, M.A.
HESIOD AND THEOGNIS. By the Rev. J. Davies, M.A.
GREEK ANTHOLOGY. By Lord Neaves.
VIRGIL. By the Editor.
HORACE. By Sir Theodore Martin, K.C.B.
JUVENAL. By Edward Walford, M.A.
PLAUTUS AND TERENCE. By the Editor.
THE COMMENTARIES OF CÆSAR. By Anthony Trollope.
TACITUS. By W. B. Donne.
CICERO. By the Editor.
PLINY'S LETTERS. By the Rev. Alfred Church, M.A., and the Rev. W. J. Brodribb, M.A.
LIVY. By the Editor.
OVID. By the Rev. A. Church, M.A.
CATULLUS, TIBULLUS, AND PROPERTIUS. By the Rev. Jas. Davies, M.A.
DEMOSTHENES. By the Rev. W. J. Brodribb, M.A.
ARISTOTLE. By Sir Alexander Grant, Bart., LL.D.
THUCYDIDES. By the Editor.
LUCRETIUS. By W. H. Mallock, M.A.
PINDAR. By the Rev. F. D. Morice, M.A.

AYLWARD. The Transvaal of To-day: War, Witchcraft, Sports, and Spoils in South Africa. By ALFRED AYLWARD, Commandant, Transvaal Republic; Captain (late) Lydenberg Volunteer Corps. Second Edition. Crown 8vo, with a Map, 6s.

AYTOUN. Lays of the Scottish Cavaliers, and other Poems. By W. EDMONDSTOUNE AYTOUN, D.C.L., Professor of Rhetoric and Belles-Lettres in the University of Edinburgh. Twenty-eighth Edition. Fcap. 8vo, 7s. 6d.

—— An Illustrated Edition of the Lays of the Scottish Cavaliers. From designs by Sir NOEL PATON. Small 4to, 21s., in gilt cloth.

—— Bothwell: a Poem. Third Edition. Fcap., 7s. 6d.

—— Firmilian; or, The Student of Badajoz. A Spasmodic Tragedy. Fcap., 5s.

—— Poems and Ballads of Goethe. Translated by Professor AYTOUN and Sir THEODORE MARTIN, K.C.B. Third Edition. Fcap., 6s.

—— Bon Gaultier's Book of Ballads. By the SAME. Thirteenth Edition. With Illustrations by Doyle, Leech, and Crowquill. Post 8vo, gilt edges, 8s. 6d.

—— The Ballads of Scotland. Edited by Professor AYTOUN. Fourth Edition. 2 vols. fcap. 8vo, 12s.

—— Memoir of William E. Aytoun, D.C.L. By Sir THEODORE MARTIN, K.C.B. With Portrait. Post 8vo, 12s.

BAGOT. The Art of Poetry of Horace. Free and Explanatory Translations in Prose and Verse. By the Very Rev. DANIEL BAGOT, D.D. Third Edition, Revised, printed on *papier vergé*, square 8vo, 5s.

WILLIAM BLACKWOOD AND SONS. 5

BAIRD LECTURES. The Mysteries of Christianity. By T. J. CRAWFORD, D.D., F.R.S.E., Professor of Divinity in the University of Edinburgh, &c. Being the Baird Lecture for 1874. Crown 8vo, 7s. 6d.

——— Endowed Territorial Work : Its Supreme Importance to the Church and Country. By WILLIAM SMITH, D.D., Minister of North Leith. Being the Baird Lecture for 1875. Crown 8vo, 6s.

——— Theism. By ROBERT FLINT, D.D., LL.D., Professor of Divinity in the University of Edinburgh. Being the Baird Lecture for 1876. Third Edition. Crown 8vo, 7s. 6d.

——— Anti-Theistic Theories. By the SAME. Being the Baird Lecture for 1877. Second Edition. Crown 8vo, 10s. 6d.

BATTLE OF DORKING. Reminiscences of a Volunteer. From 'Blackwood's Magazine.' Second Hundredth Thousand. 6d.

BY THE SAME AUTHOR.

The Dilemma. Cheap Edition. Crown 8vo, 6s.

BEDFORD. The Regulations of the Old Hospital of the Knights of St John at Valetta. From a Copy Printed at Rome, and preserved in the Archives of Malta; with a Translation, Introduction, and Notes Explanatory of the Hospital Work of the Order. By the Rev. W. K. R. BEDFORD, one of the Chaplains of the Order of St John in England. Royal 8vo, with Frontispiece, Plans, &c., 7s. 6d.

BESANT. Readings from Rabelais. By WALTER BESANT, M.A. In one volume, post 8vo. [*In the press.*]

BLACKIE. Lays and Legends of Ancient Greece. By JOHN STUART BLACKIE, Professor of Greek in the University of Edinburgh. Second Edition. Fcap. 8vo. 5s.

BLACKWOOD'S MAGAZINE, from Commencement in 1817 to June 1880. Nos. 1 to 776, forming 127 Volumes.

——— Index to Blackwood's Magazine. Vols. 1 to 50. 8vo, 15s.

——— Tales from Blackwood. Forming Twelve Volumes of Interesting and Amusing Railway Reading. Price One Shilling each in Paper Cover. Sold separately at all Railway Bookstalls.

They may also be had bound in cloth, 18s., and in half calf, richly gilt, 30s. or 12 volumes in 6, Roxburghe, 21s., and half red morocco, 28s.

——— Tales from Blackwood. New Series. Complete in Twenty-four Shilling Parts. Handsomely bound in 12 vols., cloth, 30s. In leather back, Roxburghe style, 37s. 6d. In half calf, gilt, 52s. 6d. In half morocco, 55s.

——— Standard Novels. Uniform in size and legibly Printed. Each Novel complete in one volume.

Florin Series, Illustrated Boards.

TOM CRINGLE'S LOG. By Michael Scott.
THE CRUISE OF THE MIDGE. By the Same.
CYRIL THORNTON. By Captain Hamilton.
ANNALS OF THE PARISH. By John Galt.
THE PROVOST, &c. By John Galt.
SIR ANDREW WYLIE. By John Galt.
THE ENTAIL. By John Galt.
MISS MOLLY. By Beatrice May Butt.
REGINALD DALTON. By J. G. Lockhart.

PEN OWEN. By Dean Hook.
ADAM BLAIR. By J. G. Lockhart.
LADY LEE'S WIDOWHOOD. By General Sir E. B. Hamley.
SALEM CHAPEL. By Mrs Oliphant.
THE PERPETUAL CURATE. By Mrs Oliphant.
MISS MARJORIBANKS. By Mrs Oliphant.
JOHN : A Love Story. By Mrs Oliphant.

Or in Cloth Boards, 2s. 6d.

Shilling Series, Illustrated Cover.

THE RECTOR, and THE DOCTOR'S FAMILY. By Mrs Oliphant.
THE LIFE OF MANSIE WAUCH. By D. M. Moir.
PENINSULAR SCENES AND SKETCHES. By F. Hardman.

SIR FRIZZLE PUMPKIN, NIGHTS AT MESS, &c.
THE SUBALTERN.
LIFE IN THE FAR WEST. By G. F. Ruxton.
VALERIUS : A Roman Story. By J. G. Lockhart.

Or in Cloth Boards, 1s. 6d.

6 LIST OF BOOKS PUBLISHED BY

BLACKMORE. The Maid of Sker. By R. D. BLACKMORE, Author of 'Lorna Doone,' &c. Eighth Edition. Crown 8vo, 7s. 6d.

BOSCOBEL TRACTS. Relating to the Escape of Charles the Second after the Battle of Worcester, and his subsequent Adventures. Edited by J. HUGHES, Esq., A.M. A New Edition, with additional Notes and Illustrations, including Communications from the Rev. R. H. BARHAM, Author of the 'Ingoldsby Legends.' 8vo, with Engravings, 16s.

BRACKENBURY. A Narrative of the Ashanti War. Prepared from the official documents, by permission of Major-General Sir Garnet Wolseley, K.C.B., K.C.M.G. By Major H. BRACKENBURY, R.A., Assistant Military Secretary to Sir Garnet Wolseley. With Maps from the latest Surveys made by the Staff of the Expedition. 2 vols. 8vo, 25s.

BROADLEY. Tunis, Past and Present. By A. M. BROADLEY. With numerous Illustrations and Maps. 2 vols. post 8vo. [*In the Press.*

BROOKE, Life of Sir James, Rajah of Sarāwak. From his Personal Papers and Correspondence. By SPENSER ST JOHN, H.M.'s Minister-Resident and Consul-General Peruvian Republic; formerly Secretary to the Rajah. With Portrait and a Map. Post 8vo, 12s. 6d.

BROUGHAM. Memoirs of the Life and Times of Henry Lord Brougham. Written by HIMSELF. 3 vols. 8vo, £2, 8s. The Volumes are sold separately, price 16s. each.

BROWN. The Forester: A Practical Treatise on the Planting, Rearing, and General Management of Forest-trees. By JAMES BROWN, Wood-Surveyor and Nurseryman. Fifth Edition, revised and enlarged. Royal 8vo, with Engravings. [*Nearly ready.*

BROWN. The Ethics of George Eliot's Works. By JOHN CROMBIE BROWN. Third Edition. Crown 8vo, 2s. 6d.

BROWN. A Manual of Botany, Anatomical and Physiological. For the Use of Students. By ROBERT BROWN, M.A., Ph.D., F.L.S., F.R.G.S. Crown 8vo, with numerous Illustrations, 12s. 6d.

BUCHAN. Introductory Text-Book of Meteorology. By ALEXANDER BUCHAN, M.A., F.R.S.E., Secretary of the Scottish Meteorological Society, &c. Crown 8vo, with 8 Coloured Charts and other Engravings, pp. 218. 4s. 6d.

BURBIDGE. Domestic Floriculture, Window Gardening, and Floral Decorations. Being practical directions for the Propagation, Culture, and Arrangement of Plants and Flowers as Domestic Ornaments. By F. W. BURBIDGE. Second Edition. Crown 8vo, with numerous Illustrations, 7s. 6d.

——— Cultivated Plants: Their Propagation and Improvement. Including Natural and Artificial Hybridisation, Raising from Seed, Cuttings, and Layers, Grafting and Budding, as applied to the Families and Genera in Cultivation. Crown 8vo, with numerous Illustrations, 12s. 6d.

BURN. Handbook of the Mechanical Arts Concerned in the Construction and Arrangement of Dwelling-Houses and other Buildings; with Practical Hints on Road-making and the Enclosing of Land. By ROBERT SCOTT BURN, Engineer. Second Edition. Crown 8vo, 6s. 6d.

BURTON. The History of Scotland: From Agricola's Invasion to the Extinction of the last Jacobite Insurrection. By JOHN HILL BURTON, D.C.L., Historiographer-Royal for Scotland. New and Enlarged Edition, 8 vols., and Index. Crown 8vo, £3, 3s.

——— History of the British Empire during the Reign of Queen Anne. In 3 vols. 8vo. 36s.

——— The Cairngorm Mountains. Crown 8vo, 3s. 6d.

——— The Scot Abroad. Second Edition. Complete in One volume. Crown 8vo, 10s. 6d.

——— The Book-Hunter. A New and Choice Edition. With a Memoir of the Author, a Portrait etched by Mr Hole, A.R.S.A., and other Illustrations. In small 4to, on hand-made paper. [*In the Press.*

BUTE. The Roman Breviary: Reformed by Order of the Holy Œcumenical Council of Trent; Published by Order of Pope St Pius V.; and Revised by Clement VIII. and Urban VIII.; together with the Offices since granted. Translated out of Latin into English by JOHN, Marquess of Bute, K.T. In 2 vols. crown 8vo, cloth boards, edges uncut. £2, 2s.

—— The Altus of St Columba. With a Prose Paraphrase and Notes. In paper cover, 2s. 6d.

BUTT. Miss Molly. By BEATRICE MAY BUTT. Cheap Edition, 2s.

—— Delicia. By the Author of 'Miss Molly.' Fourth Edition. Crown 8vo, 7s. 6d.

CAIRD. Sermons. By JOHN CAIRD, D.D., Principal of the University of Glasgow. Fourteenth Thousand. Fcap. 8vo, 5s.

—— Religion in Common Life. A Sermon preached in Crathie Church, October 14, 1855, before Her Majesty the Queen and Prince Albert. Published by Her Majesty's Command. Cheap Edition, 3d.

CAMPBELL, Life of Colin, Lord Clyde. See General SHADWELL, at page 20.

CAMPBELL. Sermons Preached before the Queen at Balmoral. By the Rev. A. A. CAMPBELL, Minister of Crathie. Published by Command of Her Majesty. Crown 8vo, 4s. 6d.

CARLYLE. Autobiography of the Rev. Dr Alexander Carlyle, Minister of Inveresk. Containing Memorials of the Men and Events of his Time. Edited by JOHN HILL BURTON. 8vo. Third Edition, with Portrait, 14s.

CARRICK. Koumiss; or, Fermented Mare's Milk: and its Uses in the Treatment and Cure of Pulmonary Consumption, and other Wasting Diseases. With an Appendix on the best Methods of Fermenting Cow's Milk. By GEORGE L. CARRICK, M.D., L.R.C.S.E. and L.R.C.P.E., Physician to the British Embassy, St Petersburg, &c. Crown 8vo, 10s. 6d.

CAUVIN. A Treasury of the English and German Languages. Compiled from the best Authors and Lexicographers in both Languages. Adapted to the Use of Schools, Students, Travellers, and Men of Business; and forming a Companion to all German-English Dictionaries. By JOSEPH CAUVIN, LL.D. & Ph.D., of the University of Göttingen, &c. Crown 8vo, 7s. 6d.

CAVE-BROWN. Lambeth Palace and its Associations. By J. CAVE-BROWN, M.A., Vicar of Detling, Kent, and for many years Curate of Lambeth Parish Church. With an Introduction by the Archbishop of Canterbury. In One volume, with Illustrations. [In the Press.

CHARTERIS. Canonicity; or, Early Testimonies to the Existence and Use of the Books of the New Testament. Based on Kirchhoffer's 'Quellensammlung.' Edited by A. H. CHARTERIS, D.D., Professor of Biblical Criticism in the University of Edinburgh. 8vo, 18s.

—— Life of the Rev. James Robertson, D.D., F.R.S.E., Professor of Divinity and Ecclesiastical History in the University of Edinburgh. By Professor CHARTERIS. With Portrait. 8vo, 10s. 6d.

CHEVELEY NOVELS, THE.
I. A MODERN MINISTER. 2 vols. bound in cloth, with Twenty-six Illustrations. 17s.
II. SAUL WEIR. 2 vols. bound in cloth. With Twelve Illustrations by F. Barnard. 16s.

CHIROL. 'Twixt Greek and Turk. By M. VALENTINE CHIROL. Post 8vo. With Frontispiece and Map, 10s. 6d.

CHURCH SERVICE SOCIETY. A Book of Common Order: Being Forms of Worship issued by the Church Service Society. Fourth Edition, 5s.

COLQUHOUN. The Moor and the Loch. Containing Minute Instructions in all Highland Sports, with Wanderings over Crag and Corrie, Flood and Fell. By JOHN COLQUHOUN. Fifth Edition, greatly enlarged. With Illustrations. 2 vols. post 8vo, 26s.

COTTERILL. The Genesis of the Church. By the Right. Rev. HENRY COTTERILL, D.D., Bishop of Edinburgh. Demy 8vo, 16s.

CRANSTOUN. The Elegies of Albius Tibullus. Translated into English Verse, with Life of the Poet, and Illustrative Notes. By JAMES CRANSTOUN, LL.D., Author of a Translation of 'Catullus.' Crown 8vo, 6s. 6d.

——— The Elegies of Sextus Propertius. Translated into English Verse, with Life of the Poet, and Illustrative Notes. Crown 8vo, 7s. 6d.

CRAWFORD. The Doctrine of Holy Scripture respecting the Atonement. By the late THOMAS J. CRAWFORD, D.D., Professor of Divinity in the University of Edinburgh. Third Edition. 8vo, 12s.

——— The Fatherhood of God, Considered in its General and Special Aspects, and particularly in relation to the Atonement, with a Review of Recent Speculations on the Subject. Third Edition, Revised and Enlarged. 8vo, 9s.

——— The Preaching of the Cross, and other Sermons. 8vo, 7s. 6d.

——— The Mysteries of Christianity; being the Baird Lecture for 1874. Crown 8vo, 7s. 6d.

CROSSE. Round about the Carpathians. By ANDREW F. CROSSE, F.C.S. 8vo, with Map of the Author's route, price 12s. 6d.

DESCARTES. The Method, Meditations, and Principles of Philosophy of Descartes. Translated from the Original French and Latin. With a New Introductory Essay, Historical and Critical, on the Cartesian Philosophy. By JOHN VEITCH, LL.D., Professor of Logic and Rhetoric in the University of Glasgow. A New Edition, being the Eighth. Price 6s. 6d.

DICKSON. Japan; being a Sketch of the History, Government, and Officers of the Empire. By WALTER DICKSON. 8vo, 15s.

DU CANE. The Odyssey of Homer, Books I.-XII. Translated into English Verse. By Sir CHARLES DU CANE, K.C.M.G. 8vo, 10s. 6d.

DUDGEON. History of the Edinburgh or Queen's Regiment Light Infantry Militia, now Third Battalion, The Royal Scots; with an Account of the Origin and Progress of the Militia, and a Brief Sketch of the old Royal Scots. By Major R. C. DUDGEON, Adjutant 3d Battalion The Royal Scots. Post 8vo, with Illustrations, 10s. 6d.

EAGLES. Essays. By the Rev. JOHN EAGLES, A.M. Oxon. Originally published in 'Blackwood's Magazine.' Post 8vo, 10s. 6d.

——— The Sketcher. Originally published in 'Blackwood's Magazine.' Post 8vo, 10s. 6d.

ELIOT. Impressions of Theophrastus Such. By GEORGE ELIOT. New and cheaper Edition. Crown 8vo, 5s.

——— Adam Bede. Illustrated Edition. 3s. 6d., cloth.

——— The Mill on the Floss. Illustrated Edition. 3s. 6d., cloth.

——— Scenes of Clerical Life. Illustrated Edition. 3s., cloth.

——— Silas Marner: The Weaver of Raveloe. Illustrated Edition. 2s. 6d., cloth.

——— Felix Holt, the Radical. Illustrated Edition. 3s. 6d., cloth.

——— Romola. With Vignette. 3s. 6d., cloth.

——— Middlemarch. Crown 8vo, 7s. 6d.

——— Daniel Deronda. Crown 8vo, 7s. 6d.

ELIOT. Works of George Eliot (Cabinet Edition). Complete and Uniform Edition, handsomely printed in a new type, 20 volumes, crown 8vo, price £5. The Volumes are also sold separately, price 5s. each, viz.:—Romola. 2 vols.—Silas Marner, The Lifted Veil, Brother Jacob. 1 vol.—Adam Bede. 2 vols.—Scenes of Clerical Life. 2 vols.—The Mill on the Floss. 2 vols.—Felix Holt. 2 vols.—Middlemarch. 3 vols.—Daniel Deronda. 3 vols.—The Spanish Gypsy. 1 vol.—Jubal, and other Poems, Old and New. 1 vol.—Theophrastus Such. 1 vol.

—— The Spanish Gypsy. Crown 8vo, 5s.

—— The Legend of Jubal, and other Poems, Old and New. New Edition. Fcap. 8vo, 5s., cloth.

—— Wise, Witty, and Tender Sayings, in Prose and Verse. Selected from the Works of GEORGE ELIOT. Fifth Edition. Fcap. 8vo, 6s.

—— The George Eliot Birthday Book. Printed on fine paper, with red border, and handsomely bound in cloth, gilt. Fcap. 8vo, cloth, 3s. 6d. And in French morocco or Russia, 5s.

ESSAYS ON SOCIAL SUBJECTS. Originally published in the 'Saturday Review.' A New Edition. First and Second Series. 2 vols. crown 8vo, 6s. each.

EWALD. The Crown and its Advisers; or, Queen, Ministers, Lords, and Commons. By ALEXANDER CHARLES EWALD, F.S.A. Crown 8vo, 5s.

THE FAITHS OF THE WORLD. A Concise History of the Great Religious Systems of the World. By various Authors. Being the St Giles' Lectures—Second Series. Complete in One Volume, Crown 8vo, 5s. Sold separately, price 4d.

FERGUSSON. The Honourable Henry Erskine, Lord Advocate for Scotland. With Notices of certain of his Kinsfolks and of his Time. Compiled from Family Papers, and other sources of Information. By LIEUTENANT-COLONEL ALEX. FERGUSSON, late of the Staff of her Majesty's Indian Army. In One volume, large 8vo. With Portraits and other Illustrations. [*Shortly*.

FERRIER. Philosophical Works of the late James F. Ferrier, B.A. Oxon., Professor of Moral Philosophy and Political Economy, St Andrews. New Edition. Edited by Sir ALEX. GRANT, Bart., D.C.L., and Professor LUSHINGTON. 3 vols. crown 8vo, 34s. 6d.

—— Institutes of Metaphysic. Third Edition. 10s. 6d.

—— Lectures on the Early Greek Philosophy. Third Edition. 10s. 6d.

—— Philosophical Remains, including the Lectures on Early Greek Philosophy. 2 vols., 24s.

FERRIER. George Eliot and Judaism. An Attempt to appreciate 'Daniel Deronda.' By Professor DAVID KAUFMANN, of the Jewish Theological Seminary, Buda-Pesth. Translated from the German by J. W. FERRIER. Second Edition. Crown 8vo, 2s. 6d.

FLINT. The Philosophy of History in Europe. Vol. I., containing the History of that Philosophy in France and Germany. By ROBERT FLINT, D.D., LL.D., Professor of Divinity, University of Edinburgh. 8vo, 15s.

—— Theism. Being the Baird Lecture for 1876. Third Edition. Crown 8vo, 7s. 6d.

—— Anti-Theistic Theories. Being the Baird Lecture for 1877. Second Edition. Crown 8vo, 10s. 6d.

FORBES. The Campaign of Garibaldi in the Two Sicilies: A Personal Narrative. By CHARLES STUART FORBES, Commander, R.N. Post 8vo, with Portraits, 12s.

LIST OF BOOKS PUBLISHED BY

FOREIGN CLASSICS FOR ENGLISH READERS. Edited by Mrs OLIPHANT. Price 2s. 6d.

Now published:—

DANTE. By the Editor.
VOLTAIRE. By General Sir E. B. Hamley, K.C.M.G.
PASCAL. By Principal Tulloch.
PETRARCH. By Henry Reeve, C.B.
GOETHE. By A. Hayward, Q.C.
MOLIÈRE. By the Editor and F. Tarver, M.A.
MONTAIGNE. By Rev. W. L. Collins, M.A.
RABELAIS. By Walter Besant, M.A.
CALDERON. By E. J. Hasell.

SAINT SIMON. By Clifton W. Collins, M.A.
CERVANTES. By the Editor.
CORNEILLE AND RACINE. By Henry M. Trollope.
MADAME DE SÉVIGNÉ. By Miss Thackeray.
LA FONTAINE, AND OTHER FRENCH FABULISTS. By Rev. W. L. Collins, M.A.
SCHILLER. By James Sime, M.A., Author of 'Lessing: his Life and Writings.'

In preparation:—ROUSSEAU. By Henry Graham.—TASSO. By E. J. Hasell.

FRASER. Handy Book of Ornamental Conifers, and of Rhododendrons and other American Flowering Shrubs, suitable for the Climate and Soils of Britain. With descriptions of the best kinds, and containing Useful Hints for their successful Cultivation. By HUGH FRASER, Fellow of the Botanical Society of Edinburgh. Crown 8vo. 6s.

GALT. Annals of the Parish. By JOHN GALT. Fcap. 8vo, 2s.

——— The Provost. Fcap. 8vo, 2s.

——— Sir Andrew Wylie. Fcap. 8vo, 2s.

——— The Entail; or, The Laird of Grippy. Fcap. 8vo, 2s.

GARDENER, THE: A Magazine of Horticulture and Floriculture. Edited by DAVID THOMSON, Author of 'The Handy Book of the Flower-Garden,' &c.; Assisted by a Staff of the best practical Writers. Published Monthly, 6d.

GENERAL ASSEMBLY OF THE CHURCH OF SCOTLAND.

——— Family Prayers. Authorised by the General Assembly of the Church of Scotland. A New Edition, crown 8vo, in large type, 4s. 6d. Another Edition, crown 8vo, 2s.

——— Prayers for Social and Family Worship. For the Use of Soldiers, Sailors, Colonists, and Sojourners in India, and other Persons, at home and abroad, who are deprived of the ordinary services of a Christian Ministry. Cheap Edition, 1s. 6d.

——— The Scottish Hymnal. Hymns for Public Worship. Published for Use in Churches by Authority of the General Assembly. Various sizes—viz.: 1. Large type, for pulpit use, cloth, 3s. 6d. 2. Longprimer type, cloth, red edges, 1s. 6d.; French morocco, 2s. 6d.; calf, 6s. 3. Bourgeois type, cloth, red edges, 1s.; French morocco, 2s. 4. Minion type, limp cloth, 6d.; French morocco, 1s. 6d. 5. School Edition, in paper cover, 2d. 6. Children's Hymnal, paper cover, 1d. No. 2, bound with the Psalms and Paraphrases, cloth, 3s.; French morocco, 4s. 6d.; calf, 7s. 6d. No. 3, bound with the Psalms and Paraphrases, cloth, 2s.; French morocco, 3s.

——— The Scottish Hymnal, with Music. Selected by the Committees on Hymns and on Psalmody. The harmonies arranged by W. H. Monk. Cloth, 1s. 6d.; French morocco, 3s. 6d. The same in the Tonic Sol-fa Notation, 1s. 6d. and 3s. 6d.

——— The Scottish Hymnal, with Fixed Tune for each Hymn. Longprimer type, 3s. 6d.

GERARD. Reata: What's in a Name? By E. D. GERARD. New Edition. In one volume, crown 8vo, 6s.

——— Beggar my Neighbour. A Novel. New Edition, complete in one volume, crown 8vo, 6s.

WILLIAM BLACKWOOD AND SONS. 11

GLEIG. The Subaltern. By G. R. GLEIG, M.A., late Chaplain-General of Her Majesty's Forces. Originally published in 'Blackwood's Magazine.' Library Edition. Revised and Corrected, with a New Preface. Crown 8vo, 7s. 6d.

GOETHE'S FAUST. Translated into English Verse by Sir THEODORE MARTIN, K.C.B. Second Edition, post 8vo, 6s. Cheap Edition, fcap., 3s. 6d.

—— Poems and Ballads of Goethe. Translated by Professor AYTOUN and Sir THEODORE MARTIN, K.C.B. Third Edition, fcap. 8vo, 6s.

GORDON CUMMING. At Home in Fiji. By C. F. GORDON CUMMING, Author of 'From the Hebrides to the Himalayas.' Fourth Edition, complete in one volume post 8vo. With Illustrations and Map. 7s. 6d.

—— A Lady's Cruise in a French Man-of-War. New and Cheaper Edition. In one volume, 8vo. With Illustrations and Map. 12s. 6d.

GRAHAM. Annals and Correspondence of the Viscount and First and Second Earls of Stair. By JOHN MURRAY GRAHAM. 2 vols. demy 8vo, with Portraits and other Illustrations. £1, 8s.

—— Memoir of Lord Lynedoch. Second Edition, crown 8vo, 5s.

GRANT. Bush-Life in Queensland. By A. C. GRANT. New Edition. In One volume crown 8vo, 6s.

GRANT. Incidents in the Sepoy War of 1857-58. Compiled from the Private Journals of the late General Sir HOPE GRANT, G.C.B.; together with some Explanatory Chapters by Captain HENRY KNOLLYS, R.A. Crown 8vo, with Map and Plans, 12s.

GRANT. Memorials of the Castle of Edinburgh. By JAMES GRANT. A New Edition. Crown 8vo, with 12 Engravings, 3s.

HAMERTON. Wenderholme: A Story of Lancashire and Yorkshire Life. By PHILIP GILBERT HAMERTON, Author of 'A Painter's Camp.' A New Edition. Crown 8vo, 6s.

HAMILTON. Lectures on Metaphysics. By Sir WILLIAM HAMILTON, Bart., Professor of Logic and Metaphysics in the University of Edinburgh. Edited by the Rev. H. L. MANSEL, B.D., LL.D., Dean of St Paul's; and JOHN VEITCH, M.A., Professor of Logic and Rhetoric, Glasgow. Sixth Edition. 2 vols. 8vo, 24s.

—— Lectures on Logic. Edited by the SAME. Third Edition. 2 vols. 24s.

—— Discussions on Philosophy and Literature, Education and University Reform. Third Edition. 8vo, 21s.

—— Memoir of Sir William Hamilton, Bart., Professor of Logic and Metaphysics in the University of Edinburgh. By Professor VEITCH of the University of Glasgow. 8vo, with Portrait, 18s.

HAMILTON. Annals of the Peninsular Campaigns. By Captain THOMAS HAMILTON. Edited by F. Hardman. 8vo, 16s. Atlas of Maps to illustrate the Campaigns, 12s.

HAMLEY. The Operations of War Explained and Illustrated. By General Sir EDWARD BRUCE HAMLEY, K.C.M.G. Fourth Edition, revised throughout. 4to, with numerous Illustrations, 30s.

—— Thomas Carlyle: An Essay. Second Edition. Crown 8vo. 2s. 6d.

—— The Story of the Campaign of Sebastopol. Written in the Camp. With Illustrations drawn in Camp by the Author. 8vo, 21s.

—— On Outposts. Second Edition. 8vo, 2s.

—— Wellington's Career; A Military and Political Summary. Crown 8vo, 2s.

—— Lady Lee's Widowhood. Crown 8vo, 2s. 6d.

HAMLEY. Our Poor Relations. A Philozoic Essay. With Illustrations, chiefly by Ernest Griset. Crown 8vo, cloth gilt, 3s. 6d.

HAMLEY. Guilty, or Not Guilty? A Tale. By Major-General W. G. HAMLEY, late of the Royal Engineers. New Edition. Crown 8vo, 3s. 6d.

—— The House of Lys : One Book of its History. A Tale. Second Edition. 2 vols. crown 8vo. 17s.

—— Traseaden Hall. "When George the Third was King." 3 vols., post 8vo, 25s. 6d.

HANDY HORSE-BOOK ; or, Practical Instructions in Riding, Driving, and the General Care and Management of Horses. By 'MAGENTA.' Ninth Edition, with 6 Engravings, 4s. 6d.

BY THE SAME.

Our Domesticated Dogs: their Treatment in reference to Food, Diseases, Habits, Punishment, Accomplishments. Crown 8vo, 2s. 6d.

HARBORD. Definitions and Diagrams in Astronomy and Navigation. By the Rev. J. B. HARBORD, M.A., Assistant Director of Education, Admiralty. 1s.

—— Short Sermons for Hospitals and Sick Seamen. Fcap. 8vo, cloth, 4s. 6d.

HARDMAN. Scenes and Adventures in Central America. Edited by FREDERICK HARDMAN. Crown 8vo, 6s.

HAWKEY. The Shakespeare Tapestry. Woven in Verse. By C. HAWKEY. Fcap. 8vo. 6s.

HAY. The Works of the Right Rev. Dr George Hay, Bishop of Edinburgh. Edited under the Supervision of the Right Rev. Bishop STRAIN. With Memoir and Portrait of the Author. 5 vols. crown 8vo, bound in extra cloth, £1, 1s. Or, sold separately—viz.:

—— The Sincere Christian Instructed in the Faith of Christ from the Written Word. 2 vols., 8s.

—— The Devout Christian Instructed in the Law of Christ from the Written Word. 2 vols., 8s.

—— The Pious Christian Instructed in the Nature and Practice of the Principal Exercises of Piety. 1 vol., 4s.

HEATLEY. The Horse-Owner's Safeguard. A Handy Medical Guide for every Man who owns a Horse. By G. S. HEATLEY, V.S. In one volume, crown 8vo. [In the Press.

HEMANS. The Poetical Works of Mrs Hemans. Copyright Editions.

One Volume, royal 8vo, 5s.

The Same, with Illustrations engraved on Steel, bound in cloth, gilt edges, 7s. 6d.

Six Volumes in Three, fcap., 12s. 6d.

SELECT POEMS OF MRS HEMANS. Fcap., cloth, gilt edges, 3s.

—— Memoir of Mrs Hemans. By her SISTER. With a Portrait, fcap. 8vo, 5s.

HOLE. A Book about Roses: How to Grow and Show Them. By the Rev. Canon HOLE. With coloured Frontispiece by the Hon. Mrs Francklin. Seventh Edition, revised. Crown 8vo, 7s. 6d.

HOMER. The Odyssey. Translated into English Verse in the Spenserian Stanza. By PHILIP STANHOPE WORSLEY. Third Edition, 2 vols., fcap., 12s.

—— The Iliad. Translated by P. S. WORSLEY and Professor CONINGTON. 2 vols. crown 8vo, 21s.

WILLIAM BLACKWOOD AND SONS. 13

HOME PRAYERS. By Ministers of the Church of Scotland and Members of the Church Service Society. Fcap. 8vo, price 3s.

HOSACK. Mary Queen of Scots and Her Accusers. Containing a Variety of Documents never before published. By JOHN HOSACK, Barrister-at-Law. A New and Enlarged Edition, with a Photograph from the Bust on the Tomb in Westminster Abbey. 2 vols. 8vo, £1, 1s.

INDEX GEOGRAPHICUS: Being a List, alphabetically arranged, of the Principal Places on the Globe, with the Countries and Subdivisions of the Countries in which they are situated, and their Latitudes and Longitudes. Applicable to all Modern Atlases and Maps. Imperial 8vo, pp. 676, 21s.

JAMIESON. The Laird's Secret. By J. H. JAMIESON. In 2 vols., crown 8vo.

JEAN JAMBON. Our Trip to Blunderland; or, Grand Excursion to Blundertown and Back. By JEAN JAMBON. With Sixty Illustrations designed by CHARLES DOYLE, engraved by DALZIEL. Fourth Thousand. Handsomely bound in cloth, gilt edges, 6s. 6d. Cheap Edition, cloth, 3s. 6d. In boards, 2s. 6d.

JOHNSON. The Scots Musical Museum. Consisting of upwards of Six Hundred Songs, with proper Basses for the Pianoforte. Originally published by JAMES JOHNSON; and now accompanied with Copious Notes and Illustrations of the Lyric Poetry and Music of Scotland, by the late WILLIAM STENHOUSE; with additional Notes and Illustrations, by DAVID LAING and C. K. SHARPE. 4 vols. 8vo, Roxburghe binding, £2, 12s. 6d.

JOHNSTON. The Chemistry of Common Life. By Professor J. F. W. JOHNSTON. New Edition, Revised, and brought down to date. By ARTHUR HERBERT CHURCH, M.A. Oxon.; Author of 'Food: its Sources, Constituents, and Uses;' 'The Laboratory Guide for Agricultural Students;' 'Plain Words about Water,' &c. Illustrated with Maps and 102 Engravings on Wood. Complete in One Volume, crown 8vo, pp. 618, 7s. 6d.

—— Professor Johnston's Elements of Agricultural Chemistry and Geology. Twelfth Edition, Revised, and brought down to date. By CHARLES A. CAMERON, M.D., F.R.C.S.I., &c. Fcap. 8vo, 6s. 6d.

—— Catechism of Agricultural Chemistry and Geology. An entirely New Edition, revised and enlarged, by CHARLES A. CAMERON, M.D., F.R.C.S.I., &c. Eighty-first Thousand, with numerous Illustrations, 1s.

JOHNSTON. Patrick Hamilton: a Tragedy of the Reformation in Scotland, 1528. By J. P. JOHNSTON. Crown 8vo, with Two Etchings by the Author, 5s.

KEITH ELPHINSTONE. Memoir of the Honourable George Keith Elphinstone, K.B., Viscount Keith of Stonehaven Marischal, Admiral of the Red.—*See* ALEXANDER ALLARDYCE, at page 4.

KING. The Metamorphoses of Ovid. Translated in English Blank Verse. By HENRY KING, M.A., Fellow of Wadham College, Oxford, and of the Inner Temple, Barrister-at-Law. Crown 8vo, 10s. 6d.

KINGLAKE. History of the Invasion of the Crimea. By A. W. KINGLAKE. Cabinet Edition. Six Volumes, crown 8vo, at 6s. each. The Volumes respectively contain:—
 I. THE ORIGIN OF THE WAR between the Czar and the Sultan.
 II. RUSSIA MET AND INVADED. With 4 Maps and Plans.
 III. THE BATTLE OF THE ALMA. With 14 Maps and Plans.
 IV. SEBASTOPOL AT BAY. With 10 Maps and Plans.
 V. THE BATTLE OF BALACLAVA. With 10 Maps and Plans.
 VI. THE BATTLE OF INKERMAN. With 11 Maps and Plans.
 VII. WINTER TROUBLES. With Map. [*In the Press.*

—— History of the Invasion of the Crimea. Vol. VI. Winter Troubles. Demy 8vo, with a Map, 16s.

—— Eothen. A New Edition, uniform with the Cabinet Edition of the 'History of the Crimean War,' price 6s.

KNOLLYS. The Elements of Field-Artillery. Designed for the Use of Infantry and Cavalry Officers. By HENRY KNOLLYS, Captain Royal Artillery; Author of 'From Sedan to Saarbrück,' Editor of 'Incidents in the Sepoy War,' &c. With Engravings. Crown 8vo, 7s. 6d.

LAKEMAN. What I saw in Kaffir-land. By Sir STEPHEN LAKEMAN (MAZHAR PACHA). Post 8vo, 8s. 6d.

LAVERGNE. The Rural Economy of England, Scotland, and Ireland. By LEONCE DE LAVERGNE. Translated from the French. With Notes by a Scottish Farmer. 8vo, 12s.

LEE. Lectures on the History of the Church of Scotland, from the Reformation to the Revolution Settlement. By the late Very Rev. JOHN LEE, D.D., LL.D., Principal of the University of Edinburgh. With Notes and Appendices from the Author's Papers. Edited by the Rev. WILLIAM LEE, D.D. 2 vols. 8vo, 21s.

LEE-HAMILTON. Poems and Transcripts. By EUGENE LEE-HAMILTON. Crown 8vo, 6s.

LEWES. The Physiology of Common Life. By GEORGE H. LEWES, Author of 'Sea-side Studies,' &c. Illustrated with numerous Engravings. 2 vols., 12s.

LOCKHART. Doubles and Quits. By Laurence W. M. LOCKHART. With Twelve Illustrations. Third Edition. Crown 8vo, 6s.

—— Fair to See: a Novel. Seventh Edition, crown 8vo, 6s.

—— Mine is Thine: a Novel. Seventh Edition, crown 8vo, 6s.

LORIMER. The Institutes of Law: A Treatise of the Principles of Jurisprudence as determined by Nature. By JAMES LORIMER, Regius Professor of Public Law and of the Law of Nature and Nations in the University of Edinburgh. New Edition, revised throughout, and much enlarged. 8vo, 18s.

—— The Institutes of the Law of Nations. A Treatise of the Jural Relation of Separate Political Communities. 8vo. [In the Press.

LYON. History of the Rise and Progress of Freemasonry in Scotland. By DAVID MURRAY LYON, Secretary to the Grand Lodge of Scotland. In small quarto. Illustrated with numerous Portraits of Eminent Members of the Craft, and Facsimiles of Ancient Charters and other Curious Documents. £1, 11s. 6d.

M'COMBIE. Cattle and Cattle-Breeders. By WILLIAM M'COMBIE, Tillyfour. A New and Cheaper Edition, 2s. 6d., cloth.

MACRAE. A Handbook of Deer-Stalking. By ALEXANDER MACRAE, late Forester to Lord Henry Bentinck. With Introduction by HORATIO ROSS, Esq. Fcap. 8vo, with two Photographs from Life. 3s. 6d.

M'CRIE. Works of the Rev. Thomas M'Crie, D.D. Uniform Edition. Four vols. crown 8vo, 24s.

—— Life of John Knox. Containing Illustrations of the History of the Reformation in Scotland. Crown 8vo, 6s. Another Edition, 3s. 6d.

—— Life of Andrew Melville. Containing Illustrations of the Ecclesiastical and Literary History of Scotland in the Sixteenth and Seventeenth Centuries. Crown 8vo, 6s.

—— History of the Progress and Suppression of the Reformation in Italy in the Sixteenth Century. Crown 8vo, 4s.

—— History of the Progress and Suppression of the Reformation in Spain in the Sixteenth Century. Crown 8vo, 3s. 6d.

—— Sermons, and Review of the 'Tales of My Landlord.' Crown 8vo, 6s.

—— Lectures on the Book of Esther. Fcap. 8vo, 5s.

WILLIAM BLACKWOOD AND SONS. 15

M'INTOSH. The Book of the Garden. By CHARLES M'INTOSH, formerly Curator of the Royal Gardens of his Majesty the King of the Belgians, and lately of those of his Grace the Duke of Buccleuch, K.G., at Dalkeith Palace. Two large vols. royal 8vo, embellished with 1350 Engravings. £4, 7s. 6d.
Vol. I. On the Formation of Gardens and Construction of Garden Edifices. 776 pages, and 1073 Engravings, £2, 10s.
Vol. II. Practical Gardening. 868 pages, and 279 Engravings, £1, 17s. 6d.

MACKAY. A Manual of Modern Geography; Mathematical, Physical, and Political. By the Rev. ALEXANDER MACKAY, LL.D., F.R.G.S. New and Greatly Improved Edition. Crown 8vo, pp. 688. 7s. 6d.

—— Elements of Modern Geography. 47th Thousand, revised to the present time. Crown 8vo, pp. 300, 3s.

—— The Intermediate Geography. Intended as an Intermediate Book between the Author's 'Outlines of Geography,' and 'Elements of Geography.' Eighth Edition, crown 8vo, pp. 224, 2s.

—— Outlines of Modern Geography. 142d Thousand, revised to the Present Time. 18mo, pp. 112, 1s.

—— First Steps in Geography. 82d Thousand. 18mo, pp. 56. Sewed, 4d.; cloth, 6d.

—— Elements of Physiography and Physical Geography. With Express Reference to the Instructions recently issued by the Science and Art Department. 19th Thousand. Crown 8vo, 1s. 6d.

—— Facts and Dates; or, the Leading Events in Sacred and Profane History, and the Principal Facts in the various Physical Sciences. The Memory being aided throughout by a Simple and Natural Method. For Schools and Private Reference. New Edition, thoroughly Revised. Crown 8vo, 3s. 6d.

MACKENZIE. Studies in Roman Law. With Comparative Views of the Laws of France, England, and Scotland. By LORD MACKENZIE, one of the Judges of the Court of Session in Scotland. Fifth Edition, Edited by JOHN KIRKPATRICK, Esq., M.A. Cantab.; Dr Jur. Heidelb.; LL.B., Edin.; Advocate. 8vo, 12s.

MANNERS. Notes of an Irish Tour in 1846. By Lord JOHN MANNERS, M.P., G.C.B. New Edition, crown 8vo. 2s. 6d.

MARMORNE. The Story is told by ADOLPHUS SEGRAVE, the youngest of three Brothers. Third Edition. Crown 8vo, 6s.

MARSHALL. French Home Life. By FREDERIC MARSHALL.
CONTENTS: Servants.— Children.— Furniture.—Food.— Manners.— Language.—Dress.—Marriage. Second Edition. 5s.

MARSHMAN. History of India. From the Earliest Period to the Close of the India Company's Government; with an Epitome of Subsequent Events. By JOHN CLARK MARSHMAN, C.S.I. Abridged from the Author's larger work. New Edition, revised. Crown 8vo, with Map, 6s. 6d.

MARTIN. Goethe's Faust. Translated by Sir THEODORE MARTIN, K.C.B. Second Edition, crown 8vo, 6s. Cheap Edition, 3s. 6d.

—— The Works of Horace. Translated into English Verse, with Life and Notes. In 2 vols. crown 8vo, printed on hand-made paper, 21s.

—— Poems and Ballads of Heinrich Heine. Done into English Verse. Second Edition. Printed on papier vergé, crown 8vo, 8s.

—— Catullus. With Life and Notes. Second Edition, post 8vo, 7s. 6d.

—— The Vita Nuova of Dante. With an Introduction and Notes. Second Edition, crown 8vo, 5s.

—— Aladdin: A Dramatic Poem. By ADAM OEHLENSCHLAEGER. Fcap. 8vo, 5s.

MARTIN. Correggio: A Tragedy. By OEHLENSCHLAEGER. With Notes. Fcap. 8vo, 3s.

———— King Rene's Daughter: A Danish Lyrical Drama. By HENRIK HERTZ. Second Edition, fcap., 2s. 6d.

MEIKLEJOHN. An Old Educational Reformer—Dr Bell. By J. M. D. MEIKLEJOHN, M.A., Professor of the Theory, History, and Practice of Education in the University of St Andrews. Crown 8vo, 3s. 6d.

MICHEL. A Critical Inquiry into the Scottish Language. With the view of Illustrating the Rise and Progress of Civilisation in Scotland. By FRANCISQUE-MICHEL, F.S.A. Lond. and Scot., Correspondant de l'Institut de France, &c. In One handsome Quarto Volume, printed on hand-made paper, and appropriately bound in Roxburghe style. Price 66s. *The Edition is strictly limited to 500 copies, which will be numbered and allotted in the order of application.*

MICHIE. The Larch: Being a Practical Treatise on its Culture and General Management. By CHRISTOPHER YOUNG MICHIE, Forester, Cullen House. Crown 8vo, with Illustrations. 7s. 6d.

MINTO. A Manual of English Prose Literature, Biographical and Critical: designed mainly to show Characteristics of Style. By W. MINTO, M.A., Professor of Logic in the University of Aberdeen. Second Edition, revised. Crown 8vo, 7s. 6d.

———— Characteristics of English Poets, from Chaucer to Shirley. Crown 8vo, 9s.

MITCHELL. Biographies of Eminent Soldiers of the last Four Centuries. By Major-General JOHN MITCHELL, Author of 'Life of Wallenstein.' With a Memoir of the Author. 8vo, 9s.

MOIR. Poetical Works of D. M. MOIR (Delta). With Memoir by THOMAS AIRD, and Portrait. Second Edition, 2 vols. fcap. 8vo, 12s.

———— Domestic Verses. New Edition, fcap. 8vo, cloth gilt, 4s. 6d.

———— Lectures on the Poetical Literature of the Past Half-Century. Third Edition, fcap. 8vo, 5s.

———— Life of Mansie Wauch, Tailor in Dalkeith. With 8 Illustrations on Steel, by the late GEORGE CRUIKSHANK. Crown 8vo. 3s. 6d. Another Edition, fcap. 8vo, 1s. 6d.

MOMERIE. The Origin of Evil; and other Sermons. Preached in St Peter's, Cranley Gardens. By the Rev. A. W. MOMERIE, M.A., D.Sc., Fellow of St John's College, Cambridge; Professor of Logic and Metaphysics in King's College, London. Second Edition, enlarged. Crown 8vo, 5s.

———— Personality. The Beginning and End of Metaphysics, and the Necessary Assumption in all Positive Philosophy. Crown 8vo, 3s.

MONTAGUE. Campaigning in South Africa. Reminiscences of an Officer in 1879. By Captain W. E. MONTAGUE, 94th Regiment, Author of 'Claude Meadowleigh,' &c. 8vo, 10s. 6d.

MONTALEMBERT. Count de Montalembert's History of the Monks of the West. From St Benedict to St Bernard. Translated by Mrs OLIPHANT. 7 vols. 8vo, £3, 17s. 6d.

———— Memoir of Count de Montalembert. A Chapter of Recent French History. By Mrs OLIPHANT, Author of the 'Life of Edward Irving,' &c. 2 vols. crown 8vo, £1, 4s.

MORE THAN KIN. By M. P. One volume, crown 8vo, 7s. 6d.

MURDOCH. Manual of the Law of Insolvency and Bankruptcy: Comprehending a Summary of the Law of Insolvency, Notour Bankruptcy, Composition-contracts, Trust-deeds, Cessios, and Sequestrations; and the Winding-up of Joint-Stock Companies in Scotland; with Annotations on the various Insolvency and Bankruptcy Statutes; and with Forms of Procedure applicable to these Subjects. By JAMES MURDOCH, Member of the Faculty of Procurators in Glasgow. Fourth Edition, Revised and Enlarged, 8vo, £1.

NEAVES. A Glance at some of the Principles of Comparative Philology. As illustrated in the Latin and Anglican Forms of Speech. By the Hon. Lord NEAVES. Crown 8vo, 1s. 6d.

───── Songs and Verses, Social and Scientific. By an Old Contributor to 'Maga.' Fifth Edition, fcap. 8vo, 4s.

───── The Greek Anthology. Being Vol. XX. of 'Ancient Classics for English Readers.' Crown 8vo, 2s. 6d.

NEW VIRGINIANS, THE. By the Author of 'Estelle Russell,' 'Junia,' &c. In 2 vols., post 8vo, 18s.

NICHOLSON. A Manual of Zoology, for the Use of Students. With a General Introduction on the Principles of Zoology. By HENRY ALLEYNE NICHOLSON, M.D., F.R.S.E., F.G.S., &c., Professor of Natural History in the University of Aberdeen. Sixth Edition, revised and enlarged. Crown 8vo, pp. 866, with 452 Engravings on Wood, 14s.

───── Text-Book of Zoology, for the Use of Schools. Third Edition, enlarged. Crown 8vo, with 225 Engravings on Wood, 6s.

───── Introductory Text-Book of Zoology, for the Use of Junior Classes. Third Edition, revised and enlarged, with 136 Engravings, 3s.

───── Outlines of Natural History, for Beginners; being Descriptions of a Progressive Series of Zoological Types. Second Edition, with Engravings, 1s. 6d.

───── A Manual of Palæontology, for the Use of Students. With a General Introduction on the Principles of Palæontology. Second Edition. Revised and greatly enlarged. 2 vols. 8vo, with 722 Engravings, £2, 2s.

───── The Ancient Life-History of the Earth. An Outline of the Principles and Leading Facts of Palæontological Science. Crown 8vo, with numerous Engravings, 10s. 6d.

───── On the "Tabulate Corals" of the Palæozoic Period, with Critical Descriptions of Illustrative Species. Illustrated with 15 Lithograph Plates and numerous Engravings. Super-royal 8vo, 21s.

───── On the Structure and Affinities of the Genus Monticulipora and its Sub-Genera, with Critical Descriptions of Illustrative Species. Illustrated with numerous Engravings on wood and lithographed Plates. Super-royal 8vo, 18s.

───── Synopsis of the Classification of the Animal Kingdom. In one volume 8vo, with numerous Illustrations.

NICHOLSON. Communion with Heaven, and other Sermons. By the late MAXWELL NICHOLSON, D.D., Minister of St Stephen's, Edinburgh. Crown 8vo, 5s. 6d.

───── Rest in Jesus. Sixth Edition. Fcap. 8vo, 4s. 6d.

OLIPHANT. The Land of Gilead. With Excursions in the Lebanon. By LAURENCE OLIPHANT, Author of 'Lord Elgin's Mission to China and Japan,' &c. With Illustrations and Maps. Demy 8vo, 21s.

───── The Land of Khemi. Post 8vo, with Illustrations, 10s. 6d.

───── Traits and Travesties; Social and Political. Post 8vo, 10s. 6d.

───── Piccadilly: A Fragment of Contemporary Biography. With Eight Illustrations by Richard Doyle. Fifth Edition, 4s. 6d. Cheap Edition, in paper cover, 2s. 6d.

OLIPHANT. Historical Sketches of the Reign of George Second. By Mrs OLIPHANT. Third Edition, 6s.

───── The Story of Valentine; and his Brother. 5s., cloth.

───── Katie Stewart. 2s. 6d.

OLIPHANT. Salem Chapel. 2s. 6d., cloth.
—— The Perpetual Curate. 2s. 6d., cloth.
—— Miss Marjoribanks. 2s. 6d., cloth.
—— The Rector, and the Doctor's Family. 1s. 6d., cloth.
—— John : A Love Story. 2s. 6d., cloth.
OSBORN. Narratives of Voyage and Adventure. By Admiral SHERARD OSBORN, C.B. 3 vols. crown 8vo, 12s.
OSSIAN. The Poems of Ossian in the Original Gaelic. With a Literal Translation into English, and a Dissertation on the Authenticity of the Poems. By the Rev. ARCHIBALD CLERK. 2 vols. imperial 8vo, £1, 11s. 6d.
OSWALD. By Fell and Fjord ; or, Scenes and Studies in Iceland. By E. J. OSWALD. One Volume, post 8vo, with Illustrations. [*In the Press.*
PAGE. Introductory Text-Book of Geology. By DAVID PAGE, LL.D., Professor of Geology in the Durham University of Physical Science, Newcastle. With Engravings on Wood and Glossarial Index. Eleventh Edition, 2s. 6d.
—— Advanced Text-Book of Geology, Descriptive and Industrial. With Engravings, and Glossary of Scientific Terms. Sixth Edition, revised and enlarged, 7s. 6d.
—— Geology for General Readers. A Series of Popular Sketches in Geology and Palæontology. Third Edition, enlarged, 6s.
—— Chips and Chapters. A Book for Amateurs and Young Geologists. 5s.
—— The Past and Present Life of the Globe. With numerous Illustrations. Crown 8vo, 6s.
—— The Crust of the Earth : A Handy Outline of Geology. Sixth Edition, 1s.
—— Economic Geology ; or, Geology in its relation to the Arts and Manufactures. With Engravings, and Coloured Map of the British Islands. Crown 8vo, 7s. 6d.
—— Introductory Text-Book of Physical Geography. With Sketch-Maps and Illustrations. Edited by CHARLES LAPWORTH, F.G.S., &c., Professor of Geology and Mineralogy in the Mason Science College, Birmingham. 10th Edition. 2s. 6d.
—— Advanced Text-Book of Physical Geography. Second Edition. With Engravings. 5s.
PAGET. Paradoxes and Puzzles : Historical, Judicial, and Literary. Now for the first time published in Collected Form. By JOHN PAGET, Barrister-at-Law. 8vo, 12s.
PATON. Spindrift. By Sir J. NOEL PATON. Fcap., cloth, 5s.
—— Poems by a Painter. Fcap., cloth, 5s.
PATTERSON. Essays in History and Art. By R. H. PATTERSON. 8vo, 12s.
PAUL. History of the Royal Company of Archers, the Queen's Body-Guard for Scotland. By JAMES BALFOUR PAUL, Advocate of the Scottish Bar. Crown 4to, with Portraits and other Illustrations. £2, 2s.
PAUL. Analysis and Critical Interpretation of the Hebrew Text of the Book of Genesis. Preceded by a Hebrew Grammar, and Dissertations on the Genuineness of the Pentateuch, and on the Structure of the Hebrew Language. By the Rev. WILLIAM PAUL, A.M. 8vo, 18s.
PETTIGREW. The Handy Book of Bees, and their Profitable Management. By A. PETTIGREW. Fourth Edition, Enlarged, with Engravings. Crown 8vo, 3s. 6d.

PHILLIMORE. Uncle Z. By GREVILLE PHILLIMORE, Rector of Henley-on-Thames. Crown 8vo, 7s. 6d.

PHILOSOPHICAL CLASSICS FOR ENGLISH READERS. Companion Series to Ancient and Foreign Classics for English Readers. Edited by WILLIAM KNIGHT, LL.D., Professor of Moral Philosophy, University of St Andrews. In crown 8vo volumes, with portraits, price 3s. 6d.

1. DESCARTES. By Professor Mahaffy, Dublin.
2. BUTLER. By the Rev. W. Lucas Collins, M.A., Honorary Canon of Peterborough.
3. BERKELEY. By Professor A. Campbell Fraser, Edinburgh.
4. FICHTE. By Professor Adamson, Owens College, Manchester.
5. KANT. By William Wallace, M.A., LL.D., Merton College, Oxford.

POLLOK. The Course of Time: A Poem. By ROBERT POLLOK, A.M. Small fcap. 8vo, cloth gilt, 2s. 6d. The Cottage Edition, 32mo, sewed, 8d. The Same, cloth, gilt edges, 1s. 6d. Another Edition, with Illustrations by Birket Foster and others, fcap., gilt cloth, 3s. 6d., or with edges gilt, 4s.

PORT ROYAL LOGIC. Translated from the French: with Introduction, Notes, and Appendix. By THOMAS SPENCER BAYNES, LL.D., Professor in the University of St Andrews. Eighth Edition, 12mo, 4s.

POST-MORTEM. Third Edition, 1s.

BY THE SAME AUTHOR.

The Autobiography of Thomas Allen. 3 vols. post 8vo, 25s. 6d.

POTTS AND DARNELL. Aditus Faciliores: An easy Latin Construing Book, with Complete Vocabulary. By A. W. POTTS, M.A., LL.D., Head-Master of the Fettes College, Edinburgh, and sometime Fellow of St John's College, Cambridge; and the Rev. C. DARNELL, M.A., Head-Master of Cargilfield Preparatory School, Edinburgh, and late Scholar of Pembroke and Downing Colleges, Cambridge. Seventh Edition, fcap. 8vo, 3s. 6d.

——— Aditus Faciliores Graeci. An easy Greek Construing Book, with Complete Vocabulary. Third Edition, fcap. 8vo, 3s.

PRINGLE. The Live-Stock of the Farm. By ROBERT O. PRINGLE. Third Edition, crown 8vo. [In the press.

PRIVATE SECRETARY. 3 vols. post 8vo, 25s. 6d.

PUBLIC GENERAL STATUTES AFFECTING SCOTLAND, from 1707 to 1847, with Chronological Table and Index. 3 vols. large 8vo, £3, 3s.

PUBLIC GENERAL STATUTES AFFECTING SCOTLAND, COLLECTION OF. Published Annually with General Index.

RAMSAY. Rough Recollections of Military Service and Society. By Lieut.-Col. BALCARRES D. WARDLAW RAMSAY. Two vols. post 8vo.

RANKINE. A Treatise on the Rights and Burdens Incident to the Ownership of Lands and other Heritages in Scotland. By JOHN RANKINE, M.A., Advocate. Large 8vo, 40s.

REID. A Handy Manual of German Literature. By M. F. REID. For Schools, Civil Service Competitions, and University Local Examinations. Fcap. 8vo, 3s.

REVOLT OF MAN. Post 8vo, 7s. 6d.

ROBERTSON. Orellana, and other Poems. By J. LOGIE ROBERTSON. Fcap. 8vo. Printed on hand-made paper. 6s.

——— Our Holiday Among the Hills. By JAMES and JANET LOGIE ROBERTSON. Fcap. 8vo, 3s. 6d.

RUSSELL. The Haigs of Bemersyde. A Family History. By JOHN RUSSELL. Large octavo, with Illustrations. 21s.

RUSTOW. The War for the Rhine Frontier, 1870: Its Political and Military History. By Col. W. Rustow. Translated from the German, by John Layland Needham, Lieutenant R.M. Artillery. 3 vols. 8vo, with Maps and Plans, £1, 11s. 6d.

SANDFORD and TOWNSEND. The Great Governing Families of England. By J. Langton Sandford and Meredith Townsend. 2 vols. 8vo, 15s., in extra binding, with richly-gilt cover.

SCHETKY. Ninety Years of Work and Play. Sketches from the Public and Private Career of John Christian Schetky, late Marine Painter in Ordinary to the Queen. By his Daughter. Crown 8vo, 7s. 6d.

SCOTCH LOCH FISHING. By "Black Palmer." Crown 8vo, interleaved with blank pages, 4s.

SCOTTISH NATURALIST, THE. A Quarterly Magazine of Natural History. Edited by F. Buchanan White, M.D., F.L.S.; Annual Subscription, free by post, 4s.

SELLAR. Manual of the Education Acts for Scotland. By Alexander Craig Sellar, Advocate. Seventh Edition, greatly enlarged, and revised to the present time. 8vo, 15s.

SELLER and STEPHENS. Physiology at the Farm; in Aid of Rearing and Feeding the Live Stock. By William Seller, M.D., F.R.S.E., Fellow of the Royal College of Physicians, Edinburgh, formerly Lecturer on Materia Medica and Dietetics; and Henry Stephens, F.R.S.E., Author of 'The Book of the Farm,' &c. Post 8vo, with Engravings, 16s.

SETON. Memoir of Alexander Seton, Earl of Dunfermline, Seventh President of the Court of Session, and Lord Chancellor of Scotland. By George Seton, M.A. Oxon.; Author of the 'Law and Practice of Heraldry in Scotland,' &c. In 1 vol. 8vo. [*In the Press.*

SHADWELL. The Life of Colin Campbell, Lord Clyde. Illustrated by Extracts from his Diary and Correspondence. By Lieutenant-General Shadwell, C.B. 2 vols. 8vo. With Portrait, Maps, and Plans. 36s.

SIMPSON. Dogs of other Days: Nelson and Puck. By Eve Blantyre Simpson. Fcap. 8vo, with Illustrations, 4s. 6d.

SMITH. The Pastor as Preacher; or, Preaching in connection with Work in the Parish and the Study; being Lectures delivered at the Universities of Edinburgh, Aberdeen, and Glasgow. By Henry Wallis Smith, D.D., Minister of Kirknewton and East Calder; one of the Lecturers on Pastoral Theology appointed by the General Assembly of the Church of Scotland. Crown 8vo, 5s.

SMITH. Italian Irrigation: A Report on the Agricultural Canals of Piedmont and Lombardy, addressed to the Hon. the Directors of the East India Company; with an Appendix, containing a Sketch of the Irrigation System of Northern and Central India. By Lieut.-Col. R. Baird Smith, F.G.S., Captain, Bengal Engineers. Second Edition. 2 vols. 8vo, with Atlas in folio, 30s.

SMITH. Thorndale; or, The Conflict of Opinions. By William Smith, Author of 'A Discourse on Ethics,' &c. A New Edition. Crown 8vo, 10s. 6d.

—— Gravenhurst; or, Thoughts on Good and Evil. Second Edition, with Memoir of the Author. Crown 8vo, 8s.

—— A Discourse on Ethics of the School of Paley. 8vo, 4s.

—— Dramas. 1. Sir William Crichton. 2. Athelwold. 3. Guidone. 24mo, boards, 3s.

SOUTHEY. Poetical Works of Caroline Bowles Southey. Fcap. 8vo, 5s.

—— The Birthday, and other Poems. Second Edition, 5s.

—— Chapters on Churchyards. Fcap., 2s. 6d.

WILLIAM BLACKWOOD AND SONS. 21

SPEKE. What led to the Discovery of the Nile Source. By JOHN HANNING SPEKE, Captain H.M. Indian Army. 8vo, with Maps, &c., 14s.

—— Journal of the Discovery of the Source of the Nile. By J. H. SPEKE, Captain H.M. Indian Army. With a Map of Eastern Equatorial Africa by Captain SPEKE; numerous illustrations, chiefly from Drawings by Captain GRANT; and Portraits, engraved on Steel, of Captains SPEKE and GRANT. 8vo. 21s.

SPROTT. The Worship and Offices of the Church of Scotland; or, the Celebration of Public Worship, the Administration of the Sacraments, and other Divine Offices, according to the Order of the Church of Scotland. Being Lectures Delivered at the Universities of Aberdeen, Glasgow, St Andrews, and Edinburgh. By GEORGE W. SPROTT, D.D., Minister of North Berwick; one of the Lecturers on Pastoral Theology appointed by the General Assembly of the Church of Scotland. Crown 8vo, 6s.

STARFORTH. Villa Residences and Farm Architecture: A Series of Designs. By JOHN STARFORTH, Architect. 102 Engravings. Second Edition, medium 4to, £2, 17s. 6d.

STATISTICAL ACCOUNT OF SCOTLAND. Complete, with Index, 15 vols. 8vo, £16, 16s.
Each County sold separately, with Title, Index, and Map, neatly bound in cloth, forming a very valuable Manual to the Landowner, the Tenant, the Manufacturer, the Naturalist, the Tourist, &c.

STEPHENS. The Book of the Farm; detailing the Labours of the Farmer, Farm-Steward, Ploughman, Shepherd, Hedger, Farm-Labourer, Field-Worker, and Cattleman. By HENRY STEPHENS, F.R.S.E. Illustrated with Portraits of Animals painted from the life; and with 557 Engravings on Wood, representing the principal Field Operations, Implements, and Animals treated of in the Work. A New and Revised Edition, the third, in great part Rewritten. 2 vols. large 8vo, £2, 10s.

—— The Book of Farm-Buildings; their Arrangement and Construction. By HENRY STEPHENS, F.R.S.E., Author of 'The Book of the Farm;' and ROBERT SCOTT BURN. Illustrated with 1045 Plates and Engravings. Large 8vo, uniform with 'The Book of the Farm,' &c. £1, 11s. 6d.

—— The Book of Farm Implements and Machines. By J. SLIGHT and R. SCOTT BURN, Engineers. Edited by HENRY STEPHENS. Large 8vo, uniform with 'The Book of the Farm,' £2, 2s.

—— Catechism of Practical Agriculture. With Engravings. 1s.

STEWART. Advice to Purchasers of Horses. By JOHN STEWART, V.S. Author of 'Stable Economy.' 2s. 6d.

—— Stable Economy. A Treatise on the Management of Horses in relation to Stabling, Grooming, Feeding, Watering, and Working. Seventh Edition, fcap. 8vo, 6s. 6d.

STIRLING. Missing Proofs: a Pembrokeshire Tale. By M. C. STIRLING, Author of 'The Grahams of Invermoy.' 2 vols. crown 8vo, 17s.

—— The Minister's Son; or, Home with Honours. 3 vols. post 8vo. [*In the Press.*

STORMONTH. Etymological and Pronouncing Dictionary of the English Language. Including a very Copious Selection of Scientific Terms. For Use in Schools and Colleges, and as a Book of General Reference. By the Rev. JAMES STORMONTH. The Pronunciation carefully Revised by the Rev. P. H. PHELP, M.A. Cantab. Sixth Edition, with enlarged Supplement, containing many words not to be found in any other Dictionary. Crown 8vo, pp. 800. 7s. 6d.

—— The School Etymological Dictionary and Word-Book. Combining the advantages of an ordinary pronouncing School Dictionary and an Etymological Spelling-book. Fcap. 8vo, pp. 254. 2s.

STORY. Graffiti D'Italia. By W. W. STORY, Author of 'Roba di Roma.' Second Edition, fcap. 8vo, 7s. 6d.

STORY. Nero; A Historical Play. Fcap. 8vo, 6s.

——— Vallombrosa. Post 8vo, 5s.

STRICKLAND. Lives of the Queens of Scotland, and English Princesses connected with the Regal Succession of Great Britain. By AGNES STRICKLAND. With Portraits and Historical Vignettes. 8 vols. post 8vo, £4, 4s.

STURGIS. John-a-Dreams. A Tale. By JULIAN STURGIS. New Edition, crown 8vo, 3s. 6d.

——— Little Comedies, Old and New. Crown 8vo, 7s. 6d.

——— Dick's Wandering. 3 vols., post 8vo, 25s. 6d.

SUTHERLAND. Handbook of Hardy Herbaceous and Alpine Flowers, for general Garden Decoration. Containing Descriptions, in Plain Language, of upwards of 1000 Species of Ornamental Hardy Perennial and Alpine Plants, adapted to all classes of Flower-Gardens, Rockwork, and Waters; along with Concise and Plain Instructions for their Propagation and Culture. By WILLIAM SUTHERLAND, Gardener to the Earl of Minto; formerly Manager of the Herbaceous Department at Kew. Crown 8vo, 7s. 6d.

TAYLOR. Destruction and Reconstruction: Personal Experiences of the Late War in the United States. By RICHARD TAYLOR, Lieutenant-General in the Confederate Army. 8vo, 10s. 6d.

TAYLOR. The Story of My Life. By the late Colonel MEADOWS TAYLOR, Author of 'The Confessions of a Thug,' &c. &c. Edited by his Daughter. New and cheaper Edition, being the Fourth. Crown 8vo, 6s.

THOLUCK. Hours of Christian Devotion. Translated from the German of A. Tholuck, D.D., Professor of Theology in the University of Halle. By the Rev. ROBERT MENZIES, D.D. With a Preface written for this Translation by the Author. Second Edition, crown 8vo, 7s. 6d.

THOMSON. Handy Book of the Flower-Garden: being Practical Directions for the Propagation, Culture, and Arrangement of Plants in Flower-Gardens all the year round. Embracing all classes of Gardens, from the largest to the smallest. With Engraved and Coloured Plans, illustrative of the various systems of Grouping in Beds and Borders. By DAVID THOMSON, Gardener to his Grace the Duke of Buccleuch, K.G., at Drumlanrig. Third Edition, crown 8vo, 7s. 6d.

——— The Handy Book of Fruit-Culture under Glass: being a series of Elaborate Practical Treatises on the Cultivation and Forcing of Pines, Vines, Peaches, Figs, Melons, Strawberries, and Cucumbers. With Engravings of Hothouses, &c., most suitable for the Cultivation and Forcing of these Fruits. Second Edition. Crown 8vo, with Engravings, 7s. 6d.

THOMSON. A Practical Treatise on the Cultivation of the Grape-Vine. By WILLIAM THOMSON, Tweed Vineyards. Ninth Edition, 8vo, 5s.

TOM CRINGLE'S LOG. A New Edition, with Illustrations. Crown 8vo, cloth gilt, 5s. Cheap Edition, 2s.

TRAILL. Recaptured Rhymes. Being a Batch of Political and other Fugitives arrested and brought to Book. By H. D. TRAILL. Crown 8vo, 5s.

TRANSACTIONS OF THE HIGHLAND AND AGRICULTURAL SOCIETY OF SCOTLAND. Published annually, price 5s.

TROLLOPE. The Fixed Period. By ANTHONY TROLLOPE. 2 vols., fcap. 8vo, 12s.

TULLOCH. Rational Theology and Christian Philosophy in England in the Seventeenth Century. By JOHN TULLOCH, D.D., Principal of St Mary's College in the University of St Andrews; and one of her Majesty's Chaplains in Ordinary in Scotland. Second Edition. 2 vols. 8vo, 28s.

TULLOCH. Some Facts of Religion and of Life. Sermons Preached before her Majesty the Queen in Scotland, 1866-76. Second Edition, crown 8vo, 7s. 6d.

——— The Christian Doctrine of Sin; being the Croall Lecture for 1876. Crown 8vo, 6s.

——— Theism. The Witness of Reason and Nature to an All-Wise and Beneficent Creator. 8vo, 10s. 6d.

TYTLER. The Wonder-Seeker; or, The History of Charles Douglas. By M. FRASER TYTLER, Author of 'Tales of the Great and Brave,' &c. A New Edition. Fcap., 3s. 6d.

VIRGIL. The Æneid of Virgil. Translated in English Blank Verse by G. K. RICKARDS, M.A., and Lord RAVENSWORTH. 2 vols. fcap. 8vo, 10s.

WALFORD. Mr Smith: A Part of his Life. By L. B. WALFORD. Cheap Edition, 3s. 6d.

——— Pauline. Fifth Edition. Crown 8vo, 6s.

——— Cousins. Cheaper Edition. Crown 8vo, 6s.

——— Troublesome Daughters. Cheaper Edition. Crown 8vo, 6s.

——— Dick Netherby. Crown 8vo, 7s. 6d.

WARREN'S (SAMUEL) WORKS. People's Edition, 4 vols. crown 8vo, cloth, 18s. Or separately:—

Diary of a Late Physician. 3s. 6d. Illustrated, crown 8vo, 7s. 6d.

Ten Thousand A-Year. 5s.

Now and Then. The Lily and the Bee. Intellectual and Moral Development of the Present Age. 4s. 6d.

Essays: Critical, Imaginative, and Juridical. 5s.

WARREN. The Five Books of the Psalms. With Marginal Notes. By Rev. SAMUEL L. WARREN, Rector of Esher, Surrey; late Fellow, Dean, and Divinity Lecturer, Wadham College, Oxford. Crown 8vo, 5s.

WELLINGTON. Wellington Prize Essays on "the System of Field Manœuvres best adapted for enabling our Troops to meet a Continental Army." Edited by General Sir EDWARD BRUCE HAMLEY, K.C.M.G. 8vo, 12s. 6d.

WESTMINSTER ASSEMBLY. Minutes of the Westminster Assembly, while engaged in preparing their Directory for Church Government, Confession of Faith, and Catechisms (November 1644 to March 1649). Printed from Transcripts of the Originals procured by the General Assembly of the Church of Scotland. Edited by the Rev. ALEX. T. MITCHELL, D.D., Professor of Ecclesiastical History in the University of St Andrews, and the Rev. JOHN STRUTHERS, LL.D., Minister of Prestonpans. With a Historical and Critical Introduction by Professor Mitchell. 8vo, 15s.

WHITE. The Eighteen Christian Centuries. By the Rev. JAMES WHITE, Author of 'The History of France.' Seventh Edition, post 8vo, with Index, 6s.

——— History of France, from the Earliest Times. Sixth Thousand, post 8vo, with Index, 6s.

LIST OF BOOKS, ETC.

WHITE. Archæological Sketches in Scotland—Kintyre and Knapdale. By Captain T. P. WHITE, R.E., of the Ordnance Survey. With numerous Illustrations. 2 vols. folio, £4, 4s. Vol. I., Kintyre, sold separately, £2, 2s.

WILLS AND GREENE. Drawing-room Dramas for Children. By W. G. WILLS and the Hon. Mrs GREENE. Crown 8vo, 6s.

WILSON. The "Ever-Victorious Army:" A History of the Chinese Campaign under Lieut.-Col. C. G. Gordon, and of the Suppression of the Tai-ping Rebellion. By ANDREW WILSON, F.A.S.L. 8vo, with Maps, 15s.

—— The Abode of Snow: Observations on a Journey from Chinese Tibet to the Indian Caucasus, through the Upper Valleys of the Himalaya. New Edition. Crown 8vo, with Map, 10s. 6d.

WILSON. Works of Professor Wilson. Edited by his Son-in-Law, Professor FERRIER. 12 vols. crown 8vo, £2, 8s.

—— Christopher in his Sporting-Jacket. 2 vols., 8s.

—— Isle of Palms, City of the Plague, and other Poems. 4s.

—— Lights and Shadows of Scottish Life, and other Tales. 4s.

—— Essays, Critical and Imaginative. 4 vols., 16s.

—— The Noctes Ambrosianæ. Complete, 4 vols., 14s.

—— The Comedy of the Noctes Ambrosianæ. By CHRISTOPHER NORTH. Edited by JOHN SKELTON, Advocate. With a Portrait of Professor Wilson and of the Ettrick Shepherd, engraved on Steel. Crown 8vo, 7s. 6d.

—— Homer and his Translators, and the Greek Drama. Crown 8vo, 4s.

WINGATE. Annie Weir, and other Poems. By DAVID WINGATE. Fcap. 8vo, 5s.

—— Lily Neil. A Poem. Crown 8vo, 4s. 6d.

WORSLEY. Poems and Translations. By PHILIP STANHOPE WORSLEY, M.A. Edited by EDWARD WORSLEY. Second Edition, enlarged. Fcap. 8vo, 6s.

WYLDE. A Dreamer. By KATHARINE WYLDE. In 3 vols., post 8vo, 25s. 6d.

YOUNG. Songs of Béranger done into English Verse. By WILLIAM YOUNG. New Edition, revised. Fcap. 8vo, 4s. 6d.

YULE. Fortification: for the Use of Officers in the Army, and Readers of Military History. By Col. YULE, Bengal Engineers. 8vo, with numerous Illustrations, 10s. 6d.

www.ingramcontent.com/pod-product-compliance
Lightning Source LLC
Chambersburg PA
CBHW030247170426
43202CB00009B/653